ENERGY AND SOCIETY

Energy and Society is the first major text to provide an extensive _____ of energy issues informed by recent research on energy in the social sciences. Written in an engaging and accessible style it draws new thinking on uneven development, consumption, vulnerability and transition together to illustrate the social significance of energy systems in the global North and South. The book features case studies, examples, discussion questions, activities, recommended reading and more, to facilitate its use in teaching. *Energy and Society* deploys contemporary geographical concepts and approaches but is not narrowly disciplinary. Its critical perspective highlights connections between energy and significant socio-economic and political processes, such as globalisation, urbanisation, international development and social justice, and connects important issues that are often treated in isolation, such as resource availability, energy security, energy access and low-carbon transition.

Co-authored by leading researchers and based on current research and thinking in the social sciences, *Energy and Society* presents a distinctive geographical approach to contemporary energy issues. It is an essential resource for upper-level undergraduates and Master's students in geography, environmental studies, urban studies, energy studies and related fields.

Gavin Bridge is Professor of Economic Geography at Durham University, UK.

Stewart Barr is Professor of Geography at the University of Exeter, UK.

Stefan Bouzarovski is Professor of Geography at the University of Manchester, UK.

Michael Bradshaw is Professor of Global Energy at Warwick Business School, University of Warwick, UK.

Ed Brown is Senior Lecturer in Human Geography at Loughborough University, UK, and National Co-Coordinator of the UK Low Carbon Energy for Development Network.

Harriet Bulkeley is Professor of Geography at Durham University, UK.

Gordon Walker is Professor in the Lancaster Environment Centre, Lancaster University and Co-Director of the DEMAND Centre (Dynamics of Energy, Mobility and Demand), UK.

ENERGY AND SOCIETY

A CRITICAL PERSPECTIVE

GAVIN BRIDGE, STEWART BARR,
STEFAN BOUZAROVSKI, MICHAEL BRADSHAW,
ED BROWN, HARRIET BULKELEY,
AND GORDON WALKER

Routledge
Taylor & Francis Group

LONDON AND NEW YORK

First published 2018
by Routledge
2 Park Square, Milton Park, Abingdon, Oxon OX14 4RN

and by Routledge
711 Third Avenue, New York, NY 10017

Routledge is an imprint of the Taylor & Francis Group, an informa business

British Library Cataloguing-in-Publication Data
A catalogue record for this book is available from the British Library

Library of Congress Cataloging-in-Publication Data
Names: Bridge, Gavin, author.
Title: Energy and society : a critical perspective / Gavin Bridge [and six others].
Description: Abingdon, Oxon ; New York, NY : Routledge, 2018. | Includes bibliographical references and index.
Identifiers: LCCN 2017058198| ISBN 9780415740739 (hbk : alk. paper) | ISBN 9780415740746 (pbk : alk. paper) | ISBN 9781351019026 (ebk)
Subjects: LCSH: Power resources—Social aspects.
Classification: LCC TJ163.2 .B745 2018 | DDC 333.79—dc23
LC record available at https://lccn.loc.gov/2017058198

ISBN: 978-0-415-74073-9 (hbk)
ISBN: 978-0-415-74074-6 (pbk)
ISBN: 978-1-351-01902-6 (ebk)

Typeset in Minion and Trade Gothic
by Florence Production Ltd, Stoodleigh, Devon, UK
Printed and bound by CPI Group (UK) Ltd, Croydon, CR0 4YY

Contents

Figures

Tables

Definitions

Acknowledgements

This book is a product of collaborative thinking and writing. Its origins lie in the establishment of an Energy Geographies Working Group (now the Energy Geographies Research Group) within the Royal Geographical Society-Institute of British Geographers in 2011. We (the authors) were closely involved in that process, and the idea of developing a co-authored book that broke with the conventional 'technologies and resources' framework of most social science texts on energy emerged at that point. We are grateful for the guidance and support of Dr Catherine Souch (Head of Research and Education at the RGS-IBG) during the process of formally establishing the Working Group, and the many Fellows of the Society who lent their support to its creation. In recognition of these origins, we have signed over royalties from sales of this book to support the work of the Energy Geographies Research Group.

The process of evolving the book from outline to printed word has been enriched, sustained and often deflected by the research and teaching activities of the writing team. Various pieces of funded research have informed the book's content and we are pleased to recognise some of these here. Gavin Bridge, Stefan Bouzarovski and Michael Bradshaw gratefully acknowledge an ESRC Seminar Series grant (RES-451-26-0692) which provided a sustained opportunity early on to think about the geographies of energy transition. Gordon Walker thanks colleagues in the DEMAND Centre, the Sustainable Practices Research Group and Conditioning Demand project for their contributions to ideas and examples drawn on in various chapters of the book. He is also grateful to colleagues who enabled collaborative trips to Iceland, Chile and Taiwan which, as a consequence, feature in photos and case studies. Stefan Bouzarovski acknowledges the support of the European Research Council and the European Union's Seventh Framework Programme, under the EVALUATE Starting Grant (FP7/2007-2013/ERC Grant Agreement number 313478). Additional support for his contribution to this book was provided by the European Union's Horizon 2020 research and innovation programme under grant agreement 649724 (reflecting only the author's views) as well as the ESRC-funded research project Urban Transformations in South Africa through Co-Designing Energy Services Provision Pathways (URBATRANS) (ES/N014138/2). Initial scoping work for the book benefited from a workshop on teaching energy within geography and environmental studies supported by the

Higher Education Academy. The Energy and Environment Specialty Group of the American Association of Geographers have been fine fellow-travellers, and the opportunity to jointly organise a panel session around energy teaching and learning at the Annual Meeting of the American Association of Geographers in New York (2012) was especially valuable. We thank Andrew Mould at Routledge for his enthusiastic support for the original idea and the patience to play a long game. Egle Zigaite and Alaina Christensen at Routledge offered excellent editorial support, and Ting Baker provided high quality copy editing.

In developing the content of *Energy and Society: a critical perspective* we have enjoyed conversations and exchanges with our research collaborators, PhD students past and present, and colleagues near and far. We thank, in particular, Raihana Ferdous, Brian King, Magdalena Kuchler, Ankit Kumar and Isaak Vié for their contributions to the case studies; Elvan Arik, Hugo Bolzon, Neil Simcock, Marilyn Smith, Eric Verdeil and the NGO Practical Action for additional photographs included in the text; and Chris Orton of Durham Geography's Cartographic Unit for some of the graphics. Finally, we would like to thank the undergraduate and postgraduate students with whom we work. Interactions with our students have shaped our understanding of what it means to offer a critical perspective on energy-society relations, and our conviction of its importance if we are to achieve forms of energy transition that are socially just.

Introduction
A critical perspective on energy–society relations

Energy is conventionally defined as 'the capacity to do work'. A more inadequate understanding of the relationship between energy and society is hard to find. Energy's 'work' encompasses a breathtaking range of social practices and unintended consequences. It shapes the rhythms and patterns of daily life in the ways people work, rest and play; and it powers hospitals and armies, warms homes, and cools data. Energy concentrates commuters into cities and distributes products across the world; it lights the darkness and, deployed in entertainment and the cultural arts, it illuminates the human condition. In short, how societies across the world acquire and use energy is entangled with a broad array of social needs, relationships and desires. These same practices of energy gathering and energy use generate adverse consequences that also fall outside the 'capacity to do work'. Examples include the effects of air and water pollution on environmental and human health; the accumulation of waste products that expose future generations to economic and environmental risk; and insecurities and vulnerabilities created by limiting access to land and natural resources. The positive and negative consequences of harnessing energy are unevenly distributed and do not affect everyone equally. For example, pollution and other risks arising from energy production and consumption constrain the life chances of many, but often have much less impact on those who benefit most from access to energy. Choices about the technologies, resources and infrastructures that make up energy systems reflect the distribution of power in society, while also shaping future possibilities.

This book argues that addressing contemporary concerns about energy requires a social science perspective. This is because important questions about society's relationship to energy disappear from view when energy is reduced to the 'capacity to do work'. How and for whom can energy supplies be made affordable and secure? Who in society has the capacity and responsibility for mitigating climate change via interventions in the energy system? Who wins and who loses from 'disruptive' energy technologies like renewables and decentralised generation? Who has a voice when energy infrastructure decisions are made, and to what political possibilities do new energy technologies and infrastructures give rise? Securing reliable, affordable and environmentally sustainable energy is one of the grand challenges of the twenty-first century. Conventional approaches inherited from the natural sciences are no longer adequate for the task at hand. In response, this book offers a richer understanding of society's relationship with energy drawing upon recent work in the social sciences. It shows how energy is integral to different ways of life across the globe, and the contemporary challenges this poses for societies.

ENERGY AS A MULTI-DIMENSIONAL MATTER OF SOCIAL CONCERN

Energy has become increasingly prominent as a matter of social concern. 'Keeping the lights on', ensuring access to energy for those currently underserved or who cannot afford it, moving towards low carbon and renewable sources of energy: these and other concerns highlight a gap between the current and desired performance of energy systems. They suggest energy systems are currently 'failing' in significant ways, and that change is required to better align actual performance with social demands. While the reliability, cost and impacts of energy systems have long been an issue for engineers and planners, today concerns about energy regularly intrude on decision-making at scales from the personal to the geopolitical. Concerns about energy shape the choices we make as individual consumers and citizens. Techniques like carbon footprint labelling and energy efficiency ratings for household products, for example, fold technical issues of fuel mix and thermal efficiency into the realm of everyday consumer choices. Cross-border issues of energy investment and energy security frequently work their way onto the agenda at inter-governmental meetings like the G7 and G20; while international agreements to mitigate climate change, such as the Paris Agreement within the UN Framework Convention on Climate Change, centre on decarbonising energy systems and promoting energy efficiency. At the same time, long standing social concerns – like the alleviation of poverty and promotion of peace and security – are being re-interpreted in ways that place energy at their core. The UN Sustainable Development Goals for ending poverty and ensuring prosperity, for example, identify energy as "central to nearly every major challenge and opportunity the world faces ... Be it jobs, security, climate change, food production or increasing incomes, access to energy for all is essential" (UN 2018). In short, energy has proliferated as a matter of social concern and is now a key issue across a wide range of institutions, policy spaces and popular debates.

These social concerns about energy are not all of a piece. The public policy landscape around energy illustrates the diversity of energy-related social concerns, although the capacity to articulate concern and envision energy alternatives extends also to communities, corporations, and non-governmental organisations. A handful of different policy framings of the energy–society relationship exist simultaneously, from energy efficiency, (low carbon) energy transition and energy poverty to energy-for-development, energy justice and energy security. Each has a distinctive vision of a desirable energy future, represents different political and economic interests, and targets different actors

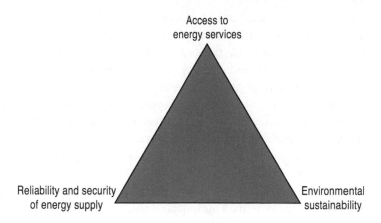

Figure I.1 The energy trilemma

and spaces of intervention. Not surprisingly, then, social concerns about energy can often drive policy in different directions: decarbonising the power sector to meet climate change goals while simultaneously promoting oil and gas extraction, for example; or encouraging local decision-making around land use planning in the name of social justice, while simultaneously centralising decisions about energy infrastructure in the interests of energy security. Energy presents society with a multi-dimensional challenge, as the design of sustainable energy systems requires the simultaneous consideration of economic, technical, political, environmental and moral questions.

The concept of an *energy trilemma* captures some of these multiple dimensions and the difficult of reconciling them (Figure I.1). The trilemma suggests public sector and private actors face the task of solving three competing social demands when it comes to the design of energy systems. First, there is the challenge of *energy security*, which reflects society's demand for reliable energy supplies and infrastructure. Second, there is the need to consider *environmental sustainability*, via efficiency of energy demand and supply and, given the dominance of high carbon sources in the global fuel mix and the urgency of climate change, a shift towards low carbon sources. And third, there is the importance of addressing *energy equity*, which encompasses issues of accessibility and affordability of supply (World Energy Council 2018).

The energy trilemma framework is useful because it distils the energy challenge into three core issues and indicates how each of these can pull policymakers and others in different directions. However, the idea that a 'happy balance' can be found, somewhere between the three poles of energy security, energy equity and environmental sustainability, is overly simplistic. In practice, the three goals are not independent of existing social structures, such as the regulatory policies of national and local governments, the investment strategies of firms, and the expectations and behaviours of energy consumers. Movement towards any of the three goals in response to the energy trilemma implies a change from the status quo, and will create both winners and losers. How these are determined – i.e. who wins and who loses – has a lot to do with the prevailing distribution of economic and political power in society. It also has a lot to do with values and ethics, as some things – like the protection of human health or the preservation of ecosystem integrity – are absolutes and cannot be reduced to trade-offs. Societies respond to the energy trilemma, then, not by coolly calibrating an optimal balance point of competing demands, but via the messy processes of political economy, which involve collaboration, conflict and co-operation. A social science approach to energy recognises this socially embedded character of the energy challenge, and how knowledge, innovative capacity, and social power are unequally distributed.

In the remaining part of this section we identify *four imperatives* shaping the evolution of energy systems. This builds on the three-sided energy trilemma outlined above but, importantly, adds a critical fourth dimension: that of social justice, which is centrally concerned with how and by whom energy systems are governed. Accordingly, we offer the 'Energy–Society Prism' (Figure I.2) as an alternative way of visualising the relationship among the four imperatives. Our contention is that the 'grand challenge' of energy system transformation cannot properly be addressed without attention to all four dimensions. The prism embodies the critical social science perspective we adopt in this book: it highlights the interaction among different imperatives while maintaining sight of social justice at all times. Energy systems have evolved in a wide range of contexts and so reflect different environmental, material and social conditions. As a consequence, the way in which these imperatives are expressed, and the balance among them in terms of the pressures shaping the evolution of a particular system, will vary depending on context. Each of the four imperatives

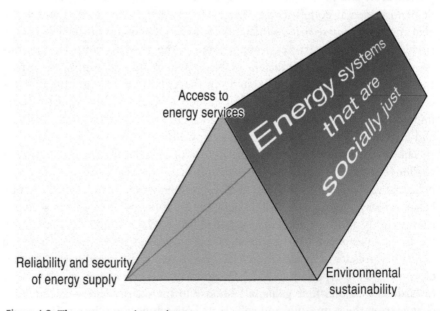

Figure I.2 The energy–society prism

is, in essence, a critique of existing energy systems. It implies energy production and consumption are failing to deliver in ways that are socially significant, and identifies the direction of desirable change. Importantly, however, the imperatives do not specify how or by whom the requisite changes should be made.

Making energy production and consumption more environmentally sustainable

Energy provides enormous economic and social advantages to those able to access it, but capturing and consuming energy can also create negative environmental effects that vary in their intensity and significance. Climate change associated with the emission of greenhouse gases, poor urban air quality, impacts to water availability and water quality, long-term management of nuclear and other energy waste streams, and competition for land use are key issues that undermine the sustainability of current energy systems. Coal, oil and gas currently provide over 80% of the world's primary energy. Unabated emissions of carbon dioxide and other greenhouse gases from the energy sector account for around two-thirds of all anthropogenic sources which are widely recognised as the primary cause of global warming. Decarbonisation of the energy sector is key to mitigating climate change, and is a process already well underway in a number of countries. The Paris Agreement within United Nations Framework Convention on Climate Change agreed a goal of sharply reducing global emissions so as to ensure a reasonable chance of "holding the increase in the global average temperature to well below 2°C above pre-industrial levels" (UNFCCC 2015). It aims to secure a near-term peak in carbon emissions and net zero emissions by the second half of the century. Reducing emissions to net zero by 2050 has substantial implications for a global energy system which is 80% dependent on high-carbon sources and where demand for modern energy services is growing. The scale of the decarbonisation challenge requires, for example, that a third of the oil reserves *currently* held by companies and governments, half of gas reserves and more than four-fifths of coal should remain in the ground (McGlade and Ekins 2016).

Environmental pressures for energy system change extend beyond global warming, however. Emissions from power stations, factories, vehicles and the use of low-quality cooking fuels like coal and wood, contribute to poor air quality conditions that are estimated to kill around 7 million people worldwide (WHO 2014). The energy sector is the main source of three pollutants responsible for widespread air quality impacts: particulate matter, sulphur oxides and nitrogen oxides (IEA 2016a). Transportation fuels, and the diesel fuel powering cars, trucks and buses in particular, is the leading source for nitrogen oxides and an important contributor to respiratory problems worldwide. In Africa, air pollution linked to energy use in transportation, electricity generation, and domestic heating and cooking is now estimated to cause more premature deaths than poor water quality or childhood malnutrition (Roy 2016). Cities are particularly effected, as "they concentrate people, energy use, construction activity and traffic . . . (and emissions discharged from transport) . . . directly into the street-level air that pedestrians breathe" (IEA 2016a: 13). The implications of growing global energy demand for water resources are increasingly recognised as a major area of concern. A primary issue is water scarcity associated with cooling at thermo-electric generating stations, the extraction and processing of fuels and the use of irrigation water for biomass production (IEA 2012, 2016b). In the UK, for example, electricity generation has been responsible for about half of all water abstractions until recently, because of the dominance of coal, gas and nuclear sources of generation. Decarbonisation pathways that involve nuclear power and carbon capture and storage are likely to intensify water demand, although those involving solar and wind power are associated with a reduction in water abstraction and consumption (Byers et al. 2014). There are also substantial water quality issues associated with the extraction and processing of fossil fuels and biofuels.

Addressing inequalities in access to modern energy services

Around the world, an estimated 1.1 billion people lack access to electricity, and 3 billion people (around nine times as many living in the United States) depend on wood, charcoal or animal waste for cooking (IEA and World Bank 2017). The number of people without access to electricity has improved slowly over time despite overall population growth; however, the number of people without access to modern cooking fuels has grown. Inequalities in energy access like these are a major cause of social injustice. The consequences of limited or zero access include poor health and reduced livelihood chances, and much of the burden of collecting traditional fuels and cooking falls disproportionately on women and children. Improving access to modern energy services, therefore, can contribute to the empowerment of women in ways that have wider social benefits, such as enhanced labour productivity and falling fertility rates. It is for this reason that improving energy access is considered a "golden thread" able to link economic growth, social equity and environmental sustainability (IEA and World Bank 2017). The goal of securing universal access is now a cornerstone of international action on energy for development. It is enshrined in the UN's Decade of Sustainable Energy for All (2014–24), for example, and as one of the UN Sustainable Development Goals: Goal 7 is to ensure universal access to modern energy services worldwide by 2030.

At the global scale, around 85% of the population have access to electricity, while around 57% are able to access 'clean cooking' fuels and technologies. There are wide disparities, however, between urban areas (which are better served) and rural areas. Both the current situation and rate of progress fall short of the rapid and far reaching transformation necessary to achieve universal access goals, which require around one billion people to receive sustainable electricity access by 2030 (Ockwell and Byrne 2016). The imperative of energy access highlights inequalities in the distribution of energy services, and draws attention to the energy services that people need in households, workplaces and communities. Using this approach, for example, the NGO Practical Action estimates half of all vaccines in the global South are ruined due to poor refrigeration, and 1 billion people use health services that have no access to electricity. In the global North, the access imperative highlights vulnerabilities and exclusions associated with energy markets, which often take geographically and historically specific forms. In the UK, for example, it is expressed in a long-running concern with affordable warmth, and the access implications of rising energy prices, falling incomes (and cuts to welfare services) and poor standards of energy efficiency in the housing stock. The NGO National Energy Action, for example, estimates 4 million UK households are affected by this 'cold homes crisis'. Similar concerns in the United States centre on the implications of rising gasoline costs (and limited public transportation) for the ability of low-paid workers to access urban job markets.

Ensuring energy systems are reliable and secure

We are accustomed to petrol being available when we need to fill up, and to lights coming on when we press the switch. But around the world, many people cannot count on a reliable and consistent supply of energy when they want it. Houses and businesses may have electrical connections, but these connections may only function for part of the day. In Nigeria, for example, 96% of households have an electrical connection but only 18% function more than half the time (Afrobarometer 2016). Limited domestic refining capacity, lack of access to foreign currency to purchase imports, and/or poor distribution networks lead to periodic gasoline and diesel fuel shortages in many parts of the global South. Technical failures and strike action can also disrupt energy supply, with some having very extensive effects: a software malfunction created an electrical blackout that lasted for two days and affected 55 million people in North America in 2003, and a blackout in northern India in July 2012 is estimated to have affected as many as 600 million people. The implications of inadequate and unreliable supply for homes and businesses are far-reaching: it undermines economic productivity, causes loss of products and data, and exacerbates inequalities in energy access as those with the means to do so to install back-up diesel generators and/or battery systems.

The imperative for reliable energy supplies requires sustained investment in energy infrastructure to enhance security of supply and build resilience to disruption when it occurs. The scale of the challenge is very large: an estimated $60 trillion dollars of energy investment is required by 2040 to meet projected increases in energy demand, with around a quarter of this going to energy efficiency (IEA 2016b). Much of the future growth in energy demand is associated with a rapidly expanding global middle class, projected to rise from 1.8 billion in 2009 to 4.9 billion by 2030 with most of this increase occurring in Asia. The projected doubling of the urban middle class, the most rapid in history and centred outside of Europe and North America, is projected to drive global economic growth through the accumulation of consumer goods and demand for energy services. This unprecedented globalisation of energy demand is having a major impact on patterns of energy trade. For countries that import a large proportion of their energy supply, guaranteeing secure and affordable supplies of energy in this context is increasingly challenging (Bradshaw 2013).

Determining how and by whom energy systems should be owned and governed

A fourth imperative is to design energy systems that are socially responsive, and deliver secure, affordable and environmentally sustainable energy in ways that promote social justice. There are many different models of ownership associated with energy production networks, including community-owned systems, public ownership via municipalities or states, and private corporations owned by shareholders. The importance of energy to economic and social life means that questions about how and by whom energy systems should be owned have figured prominently in broader debates about political philosophy. Growth of decentralised forms of renewable energy, however, has spurred an interest in local ownership models, and technologies like rooftop solar merge the role of producer and consumer in ways that challenge conventional electricity generation and distribution systems built on the separation of these roles. At the same time, a decade and more of accumulated experience of energy market liberalisation in many countries has highlighted the limits of markets and investor-owned utilities for delivering the pace and scale of changes required. Dissatisfaction with the status quo is expressed in calls for greater 'energy democracy' and experimentation with alternative models of ownership more responsive to the needs of energy consumers, and able to accommodate a desire for greater inclusion and participation in decision-making about energy futures. Prominent examples are community-ownership of renewable energy assets (such as wind farms) and a renewed interest in forms of municipal ownership around electricity, gas and heat supply. Similar concerns about the right to determine energy futures, and ensure a socially-just energy transition, now inform calls for 'energy sovereignty' by indigenous groups and other communities affected by oil, gas or large hydropower projects.

WHY TAKE A CRITICAL PERSPECTIVE?

The adjective 'critical' is widely used and can mean many things. We use it here in a specific and philosophical sense, rather than its popular meaning of a negative, scathing or unsympathetic comment. Within the social sciences, a critical perspective is one that situates a problem (like energy) within its historical and cultural context. Rather than trying to eliminate social context or hold it constant because it interferes with a search for scientific truth – which is the way many sciences work – a critical perspective actively seeks to understand how social and material conditions shape the problem. The word 'critical' here implies reflexive examination of the influence social practices, institutions and structures have on the object being studied (in this case, energy). It contrasts with a technical perspective on energy systems, which separates technical processes of energy capture and conversion from their social context in order to analyse them. The aspiration of a critical perspective is, through its awareness of social structures and material practices, to understand and explain in ways that enable improvement in social and environmental conditions. In the context of this book, a critical perspective implies attention to four things.

Socio-technical interaction: the co-constitution of energy and society

Throughout this book, we explore how energy systems and social structures shape and influence each other. The language of 'co-constitution' expresses this idea of an evolving two-way relationship, in which cause and effect work both ways. Co-constitution implies that one cannot fully understand the organisation of energy systems without reference to the organisation of society, and vice-versa. It also implies an evolutionary perspective on innovation and energy transition, in which issues like path dependency and technological and cultural forms of 'lock-in' have an important role. Co-constitution stands in contrast to a technical perspective, where explanation focuses on internal relations within a technical system. Our focus is first and foremost on the interactions between energy resources, technologies and infrastructures on the one hand, and social practices, institutions and structures on the other. This book is not the first to adopt this perspective in the context of energy. There is extensive research on the combined 'socio-technical' character of energy systems which emphasises the role of non-technical issues in the development of technological systems. The value of this socio-technical perspective is that

> it make(s) visible important aspects of energy transformation that go unrecognized and unacknowledged in other analytical approaches. These include the social *processes* that stimulate and manage energy trans-formation, the social *changes* that accompany shifts in energy technologies, and the social *outcomes* that flow from the organization and operation of novel energy systems.
>
> (Miller et al. 2013: 136, italics in original)

We draw liberally on this body of work in the chapters that follow (key references are cited, and provided at the end of each chapter).

Re-thinking energy supply and demand as material practices

'Energy' is a nineteenth-century invention, born of the industrial age of iron, coal and steam. We have inherited a valuable but limiting abstraction, originally created for the purpose of comparing different materials and technologies in the context of efforts to improve the efficiency of fuel consumption in steam boilers (a key energy technology powering nineteenth-century steam ships, mill engines and locomotives). The concept of energy and its associated metrics stripped away all material differences from fuels and machines, other those relating to their capacity to do work. This made it possible to compare very different things, such as the dimensions of large and small boilers, alternative sources of fuel and hearth designs, and the cruising speeds of ships and trains. It also allowed familiar processes – like combustion, the expansion of steam, or falling water – to be understood in terms of the transformation and conservation of energy. The concept proved exceedingly useful, enabling the science of energy physics to take hold and formulation of the laws of thermodynamics (Smith 1998). However, people do not seek energy in the abstract. Instead, they seek specific energy services like thermal comfort and mobility and these are shaped in important ways by social and geographical contexts, such as the age and type of building stock, cultural norms of heating, and distances between home and work. Similarly, the stocks and flows of energy that provide these services do not exist in an abstract form, but take the shape and character of particular materials (oil, coal, wood, animal dung) and places (narrow valleys for hydroelectric dam sites, uplands for the capture and conversion of wind energy). A critical perspective involves 'rematerialising' practices of energy production and consumption by reference to the specific materials that provide society with energy and the social practices involved in capturing and using it. A goal of the critical perspective in this book is to illustrate how the material particularities of energy production and consumption are socially significant.

How geography matters

The critical perspective we adopt in this book is alive to the way energy–society interactions take different forms over space, and how energy use underpins the geographies of everyday life. The 'grand challenge' of energy is not the same everywhere, but is made up of several distinct (although often inter-related) problems. In societies where energy is abundantly available by historical standards, for example, the primary challenge is to develop more sustainable energy systems characterised by security and affordability of supply and efficient, lower-carbon sources. For communities without abundant high quality energy services, however, and where access to energy is very limited for large proportions of the population (in much of the global South, for example, but also in poor communities within the OECD), the challenge is to enhance energy access in ways that reduce poverty and improve livelihood chances, both now and in the future. Our approach acknowledges the fact of geographical difference as a first principle, rather interpreting difference as 'local colour' or variation on (deviation from) an otherwise universal theme (Bridge 2018). This primacy of spatial difference is significant, because it means that what counts as an energy problem – and who is regarded as energy actor or agent of change – depends on where and by whom the question is asked. The critical perspective developed in this book also explores how patterns of energy use underpin the spatial and material form of cities (urban morphology), the distribution of manufacturing activity at the global scale (economic globalisation), and the connections and responsibilities created between consumers and 'distant others' (politics of consumption). A critical perspective can reveal how geographical issues like space, scale, and territory are central to contemporary energy challenges.

Social power and knowledge

Finally, a critical perspective acknowledges the role of multiple different actors – states, cities, workers, firms, international governing bodies, investors, non-governmental organisations – in shaping energy systems. It recognises their different capacities and interests, and how interactions among these actors include relationships of competition and conflict as well as partnership and collaboration. A critical perspective is interested in how the uneven distribution of economic and political power influences the evolution of the energy system in significant ways. It suggests, for example, that those with limited economic and political power are often the least well-served, whether in terms of access to energy, its affordability, or the capacity to influence decisions about energy infrastructure. There are, therefore, always economic and political tensions at the heart of energy systems – around price, profit, the distribution of environmental and social costs, and access to the levers of power – *even when these systems function well in technical and managerial terms*. Based on these insights, a critical perspective seeks to understand the energy sector, and the potential for change, in ways that do not simply reproduce existing power hierarchies. In short, it seeks to cut the cake differently, thinking across conventional divides and creating new juxtapositions in order to see relationships and processes that are occluded in conventional analyses. This book looks (and, we hope, feels) different to many other energy texts, as it does not follow the customary divisions of the energy sector based on natural science or engineering approaches. We have chosen, for example, not to organise chapters around different energy technologies (nuclear, coal, gas, biofuels etc.), or allocate separate sections for supply-side and demand-side perspectives. Instead, our goal has been to connect together issues, actors and scales of analysis that are often treated in isolation, while also drawing out new themes that are consistent with the critical perspective we have outlined here. We expand further on the book's structure in the next section.

STRUCTURE OF THE BOOK

Energy and Society: a critical perspective consists of ten chapters and is divided into three Parts. Parts 1, 2 and 3 each have an introductory overview that outlines their scope and core themes, and provides a short description of the chapters they contain.

Part 1: Energy, spaces and flows provides a foundational framework that exemplifies the book's critical perspective and focus on the co-constitution of energy and society. It adopts the concept of 'energy landscapes' as a descriptive device for holding together the material practices associated with energy production and consumption, their operation across different spatial scales (household to globe), and their dynamic and geographically uneven evolution over time. The four chapters in Part 1 examine the constitutive role of energy in social life, each focusing on a different aspect of this relationship: *Resource landscapes* (**Chapter 1**), *Economic landscapes* (**Chapter 2**), *Infrastructural landscapes* (**Chapter 3**) and *Geopolitical landscapes* (**Chapter 4**).

Part 2: Securities, vulnerabilities and justice examines the socio-economic, political and environmental consequences of current forms and patterns of energy consumption. The chapters in Part 2 highlight how energy is not consumed for its own sake, but for the services energy consumption can provide (heat, light, power and mobility), and the conditions of health, well-being, livelihood and entertainment that energy services make possible. Its focus on (in)securities, vulnerabilities and justice highlights the challenges of the contemporary energy system and how it is currently 'failing' in significant ways. Each of the four chapters in Part 2 draws together issues that are often considered separately. They highlight striking inequalities in access to energy services at a range of scales while, at the same time, disrupting conventional binaries that attribute 'excess consumption' to the global North and 'energy poverty' to the global South. The chapters in Part 2 examine *Energy poverty and vulnerability* (**Chapter 5**), *Energy consumption, inefficiency and excess* (**Chapter 6**), *Energy controversies and conflicts* (**Chapter 7**), and *Energy securities* (**Chapter 8**).

Part 3: Transitions, governance and futures considers processes of socio-technical and political change in energy systems. Demands for secure, affordable and environmentally sustainable energy mean that energy transition – along multiple and often competing dimensions – is now a social objective and policy goal. Part 3 is simultaneously historic and prospective in scope. It draws on both 'transition histories' and contemporary 'transition experiments,' and considers a range of socio-technical and political imaginaries associated with energy futures. The two chapters in Part 3 focus on *Past transitions* (**Chapter 9**) and *Future transitions* (**Chapter 10**).

REFERENCES

Afrobarometer 2016. AD75: off-grid or 'off-on': lack of access, unreliable electricity supply still plague majority of Africans. Available online at http://afrobarometer. org/publications/ad75-unreliable-electricity-supply-still-plague-majority-of-africans

Bradshaw, M. 2013. *Global energy dilemmas*. Cambridge: Polity Press.

Bridge, G. 2017. The Map is Not the Territory: a sympathetic critique of energy research's spatial turn. *Energy Research and Social Science* 36: 11–20.

Byers, E., J. Hall and J. Amezaga. 2014. Electricity generation and cooling water use: UK pathways to 2050. *Global Environmental Change* 25: 16–30.

International Energy Agency 2012. *Water for energy: is energy becoming a thirstier resource?* Excerpt from the World Energy Outlook 2012. Available online at www.worldenergyoutlook.org/media/weowebsite/2012/WEO_2012_Water_Excerpt.pdf

International Energy Agency 2016a. *Energy and Air Pollution, World Energy Outlook Special Report.* IEA/OECD, Paris.

International Energy Agency 2016b. *World Energy Outlook* 2016. Paris. Available online at www.iea.org/newsroom/news/2016/november/world-energy-outlook-2016.html

International Energy Agency (IEA) and the World Bank. 2017. *Sustainable energy for all 2017 – progress toward sustainable energy (summary).* Washington, DC: World Bank.

McGlade, C. and P. Ekins. 2016. The geographical distribution of fossil fuels unused when limiting global warming to 2°C. *Nature* 517, 187–190.

Miller, C., A. Iles and C. Jones. 2013. The social dimensions of energy transitions. *Science as Culture* 22(2): 135–148.

Ockwell, D. and R. Byrne. 2016. *Sustainable energy for all: innovation, technology and pro-poor green transformations.* Abingdon and New York: Routledge.

Roy, R. 2016. *The cost of air pollution in Africa.* Paris: OECD.

Smith, C. 1998. *The science of energy: a cultural history of energy physics in Victorian Britain.* London: The Athlone Press.

UN 2018. Sustainable Development Goals: Goal 7, Ensure access to affordable, reliable, sustainable and modern energy for all. Available online at www.un.org/sustainabledevelopment/energy/

UNFCCC (2015) Conference of the parties, Twenty-first session, Paris. Adoption of the Paris Agreement. FCCC/CP/2015/L.9 Available online at http://unfccc.int/files/home/application/pdf/paris_agreement.pdf

World Energy Council. 2018. World energy trilemma. Available at www.world energy.org/

World Health Organisation. 2014. Burden of disease from household air pollution for 2012. Available online at www.who.int/phe/health_topics/outdoorair/databases/FINAL_HAP_AAP_BoD_24March2014.pdf

Energy, spaces and flows

Part 1 – Energy, spaces and flows – explores how energy systems shape social practices and structures, and vice versa. Energy and society constitute each other in important ways, and the four chapters in this opening section examine different aspects of this two-way relationship. Each unfolds a different socio-technical 'layer' of energy–society interaction, starting with natural resources (**Chapter 1**), progressing through economies (**Chapter 2**) and infrastructures (**Chapter 3**), and arriving at geopolitics (**Chapter 4**). Core concerns across chapters in Part 1 are the material, social and spatial characteristics of energy production, distribution and consumption, and the uneven distribution of resources, knowledge, and social power.

One of the main arguments in this book is that, to develop a critical perspective on energy–society relations, we need to contextualise and materialise the abstraction (*Energy*) we have inherited from the nineteenth century (see **Introduction**). Our focus on spaces and flows in Part 1 takes this perspective forward. By *Space* we mean the places, communities, regions and territories within and through which energy capture, transformation and consumption take place. Highlighting these territorial and place-based attributes is a way of 'grounding' energy production and consumption in particular geographical and material contexts. It draws attention to the interactions, dependencies and feedbacks between these processes and the locales and settings in which they occur. In addition, it acknowledges that spatial inequalities and uneven distribution are fundamental to energy–society relationships, and need to be taken seriously in any socio-technical analysis. Space, here, is a flexible term: it does not imply one particular size or scale (e.g. city, region, or nation) but can be stretched, shrunk or folded to suit. Accordingly, the chapters in Part 1 range telescopically across a full spectrum, zooming in on the intimate scale of the household and panning back out to encompass the whole planet (and everything in between). A key idea throughout the book is that energy systems' spatial forms are dynamic: whether it is the scale of electricity networks, the shape of cities, or patterns of global trade, these spatial forms evolve with changes in energy–society relations.

Our attention to *Flows* in Part 1 is a similar contextual move, and emphasises the connections, interactions and movements forged through energy production and consumption. It draws attention to the multiple ways in places may be (dis)connected via energy, including resource movements, manufacturing trade, financial flows, infrastructural connections and the diffusion and circulation of policy ideas. Highlighting flows as a form of connection reveals how intensities of interaction vary across space (some places are more connected than others) and time (the same places can become more or less connected). Talk of spaces and flows may sound philosophical, but such issues are central to many energy-related policy concerns. In the area of transport policy, for example, advocates of aviation and high-speed rail celebrate the capacity of these infrastructures for decreasing travel time and increasing the frequency and intensity of interactions between places. Critics, however, point to the increased energy consumption (and carbon emissions) associated with higher speeds; and how, by linking centres of economic power and targeting users who are already relatively well off, high speed travel tends to intensify existing social inequalities rather than reduce them. Part 1 explores issues like these across a wide range of contexts.

The chapters in Part 1 harness the concept of *landscape* as a way of bringing together energy, spaces and flows. Landscape has the advantage of being a familiar and well-understood concept: it is, perhaps, that rare example of a well-known 'common-sense' idea that may still be made to do important analytical work. As a critical concept, landscape reaches back to efforts early in the twentieth century to understand the dual influences of cultural and physical factors in 'shaping the face of the earth' (Simmons 1996). Landscape describes a configuration of natural and cultural features across a broad space, and gestures towards the history of their production and interaction. At a time when researchers sought either to prove the determining influence of the environment on human behaviour, or to assert human control over environmental processes, landscape was an attractive concept: it provided a way to acknowledge, simultaneously, the influence of social and natural phenomena, and to hold them together rather than separate them apart or reduce one to an effect of the other. In this way landscape sat – and continues to sit – somewhat awkwardly astride one of the most enduring divisions in the history of scientific thought: the divide between the physical and social sciences. Landscape may be a familiar and apparently unadventurous idea but, from the beginning, it has cut across and challenged some of the most significant boundaries that discipline conventional thought.

The chapters in Part 1 pick up this powerful idea and explore four different instances of energy landscape. We define an energy landscape as the constellation of activities and natural and socio-technical relations through which energy production and/or consumption are achieved within a given space. Energy landscapes, therefore, simultaneously combine nature and culture, action and relation, the visible and invisible, and past and future. Energy landscapes adopt many different forms, and the chapters in Part 1 examine four broad types: resource landscapes, economic landscapes, infrastructural landscapes, and geopolitical landscapes.[1] Chapters in Part 1 mobilise the concept of landscape to achieve several different things. First, as part of a contextual perspective on energy–society relations, landscape draws attention to the interactions among natural, social and technical phenomena in ways that align well with a socio-technical approach to energy systems. It acknowledges the heterogeneity of energy systems (that they are made up of materials and relations that have different properties and qualities), and the diversity of forms into which these materials and relations can be combined (Frantál et al. 2014). An energy resource landscape such as a hydroelectric dam, oil field or wind farm, for example, is a "heterogeneous category" in that it is made up of "multiple sites, scales and forms, and involves many potential forms of social relation and engagement between 'publics' and technologies" (Walker and Cass 2007: 458).

Second, the concept draws attention to the practices and processes that lie behind the production and maintenance of energy landscapes (Nadaï and van der Horst 2010). Energy landscapes are frequently durable, shaping livelihoods and social practices over long periods of time. At the same time, they are also subject to change: energy landscapes emerge, proliferate and mature; they also decline, fall out of use and are abandoned. Most often these changes are slow, although, occasionally, they can be rapid and sudden. A landscape perspective, with its emphasis on practices and processes, shows how the durability and apparent permanence of some energy landscapes is not inherent to particular fuels or forms of consumption, but arises from repeated patterns of social interaction with material infrastructures (Haarstad and Wanvik 2016). In this way energy landscapes are "a connective tissue, a highly contextualised membrane that helps society to mould and be moulded in relation to an energy system" (Castán Broto et al. 2014: 193). Third, landscape opens a door for thinking about who governs its form and use, who is able to access the energy flows it can yield, and whose interests drive energy-related decisions. From this perspective, energy resources and infrastructures are not only 'inputs' to an energy system, but also outcomes of the way such systems are socially organised. Energy landscapes, then, can be read as expressions of economic and political power. This kind of critical landscape reading can be an effective way of opening up debate on energy futures, particularly in situations where policy options appear limited and there seems to no alternative to certain courses of action.

Fourth, landscape conveys a sense of everyday surroundings that draws attention to the way energy saturates all aspects of social life. This means not only that energy landscapes are everywhere, but also that addressing contemporary energy challenges requires configuring natural and socio-technical relations differently, and in ways that will transform the value, function and significance of established and familiar landscapes. It is no surprise, then, that "landscape has become a key arena in the debate on energy policy" (Nadaï and van der Horst 2010: 143; Pasqualetti 2011). Transition towards a low-carbon economy, for example, means scrutinising landscapes dedicated to the extraction or combustion of fossil fuels for ways in which their carbon-intensive character might be foreclosed, mitigated or offset. At the same time, landscape forms that sequester and store carbon (e.g. forests and peatlands), or that provide opportunities for the generation of low-carbon energy, gain new sources of potential value and are targeted for commercial development (Bridge et al. 2013). The concept of landscape, then, can be a useful tool for thinking through some of the most challenging issues associated with transforming energy–society relations. Fifth, landscape conveys the commonplace notion of an extensive surface bearing the imprint of human activity. It points, therefore, to the environmental, economic and political consequences of energy production, transmission, storage and consumption, and the way energy systems can leave lasting traces in the social fabric and on the built and natural environment (Mitchell 2011; Huber 2013). These far-reaching consequences of the energy system are acknowledged by the chapters in Part 1 and are the primary focus of Part 2.

INTRODUCTION TO THE CHAPTERS IN PART 1

Resource landscapes (Chapter 1) examines the massive expansion of energy availability since the Industrial Revolution; the wide range of raw materials and locations that have social value because of the energy services they can provide; and the dominance of fossil fuel extraction and thermal power generation in the global energy mix. It explores how certain socio-technical characteristics of energy resources and conversion technologies – energy density, susceptibility to control, power density, carbon intensity – influence how, where and when energy is harnessed; and the economic, political and environmental consequences to which resource landscapes give rise. All energy resource landscapes have environmental impacts, and the chapter considers how recognition of a global carbon constraint now shapes resource landscapes in significant ways. The chapter highlights the contested character of energy resource landscapes, and how they express – and, in turn, shape – the exercise of economic and political power.

Economic landscapes (Chapter 2) explores the linkages between economic activity and the availability, reliability and cost of energy at a range of scales. It outlines energy's role as a vital economic input and 'factor of production,' and shows how energy availability has re-worked several economically significant spatial forms. These include the horizontal footprint and vertical skyline of cities, the globalisation of manufacturing, and the everyday experience of space and time. Chapter 2 also explores how energy systems are currently an economic 'frontline' in a century-old struggle over whether economies should be organised around private or public forms of ownership. It highlights a renewed interest in public ownership following several decades of economic liberalisation and privatisation, and the impact of renewable energy on the emergence of alternative models to the state firm and corporate utility.

Infrastructural landscapes (Chapter 3) focuses on the networks that circulate, store and transform energy at a range of scales, from localised off-grid solutions to expansive transnational systems. It explores how energy infrastructure has become embedded in material and social life in different ways; and how moments of infrastructural breakdown (pipeline ruptures, electricity blackouts) reveal their underlying spatialities and vulnerabilities. The chapter considers the qualities of permanence, obduracy, modular growth, and invisibility that characterise these networks, and shows how they co-evolve with economic, political and technological change. The chapter highlights how, in the context of climate change, energy's infrastructure landscapes both sustain and lock in high energy ways of living, as well as offering possibilities of change towards a low carbon future.

Geopolitical landscapes (Chapter 4) focuses on how energy production, distribution and consumption shape international relations, and are embedded in the geopolitics of inter- and intra-state competition, co-operation and conflict. It considers the ways in which energy has been integral to the making of political projects in the broadest sense, including the creation of the nation state and the making of the modern citizen. It examines how the geopolitics of the contemporary carbon-intensive energy system are increasingly connected to the responses of states and cities to climate change, and the process of economic globalisation. The chapter reflects on emerging geopolitical relations around low-carbon energy, and how these are shaped by the different geographies, temporalities and scalability of low-carbon energy capture and generation technologies.

NOTE

1   Energy consumption features extensively within Part 2 (particularly **Chapters 5 and 6**), and elements of it appear in **Chapters 2, 3 and 4**.

REFERENCES

Bridge, G., S. Bouzarovski, M. Bradshaw and N. Eyre. 2013. Geographies of energy transition: space, place and the low-carbon economy. *Energy Policy* 53: 331–340.

Castán Broto, V., D. Salazar and K. Adams. 2014. Communities and urban energy landscapes in Maputo, Mozambique. *People, Place & Policy* 8(3): 192–207.

Frantál, B., M.J. Pasqualetti and D. van Der Horst. 2014. New trends and challenges for energy geographies: introduction to the Special Issue. *Moravian Geographical Reports* 22(2): 2–6.

Haarstad, H. and T. Wanvik. 2016. Carbonscapes and beyond: conceptualizing the instability of oil landscapes. *Progress in Human Geography* 41(4): 432–450.

Huber, M.T., 2013. *Lifeblood: oil, freedom, and the forces of capital.* Minneapolis, MN: University of Minnesota Press.

Mitchell, T., 2011. *Carbon democracy: Political power in the age of oil.* London and New York: Verso Books.

Nadaï, A. and D. van der Horst. 2010. Introduction: landscapes of energies. *Landscape Research* 35(2): 143–155.

Pasqualetti, M. 2011. Social barriers to renewable energy landscapes. *The Geographical Review* 101(2): 201–223.

Simmons, I. G. 1996. *Changing the face of the earth: culture, environment, history.* London: Blackwells.

Walker, G. and Cass, N., 2007. Carbon reduction, 'the public' and renewable energy: engaging with socio-technical configurations. *Area* 39(4): 458–469.

Resource landscapes

Capture. Harness. Transform. Energy resource landscapes are where societies take in energy and turn it to the task of satisfying human needs and desires. Resource landscapes *capture* raw materials like wood, coal and gas that serve society as convenient (transportable, storable) carriers of energy content. They *harness* freely available geophysical phenomena like tides, solar rays, geothermal heat, falling water and the wind. And they *transform* these **primary energy sources** into forms of energy that are socially useful through a diverse range of technologies, built structures and work practices (Figure 1.1). Classic resource landscapes include coal mines, oil fields, gas wells and dams that appropriate and/or store primary energy resources; and the power stations, refineries, district heating systems, wind farms and gas-processing plants through which raw materials and geophysical phenomena are transformed into **secondary energy sources** (electricity, diesel fuel) that do useful work. The concept of a resource landscape allows us to look beyond the technologies and resources of energy conversion, and think about their relationships to other actors (workers, residents, consumers, investors,

primary energy sources
sources of energy found in nature that have not been subject to conversion or transformation process. These can be raw materials (e.g. coal, wood) or natural phenomena (e.g. wind and tides).

secondary energy sources
a secondary energy source is derived from a primary energy resource. Examples include coke, charcoal, electricity or diesel fuel. A secondary energy source (like gasoline) can often be produced from several different primary sources (such as crude oil, coal, or bituminous sands).

governments) and wider social, economic, political and environmental consequences.

Energy resource landscapes range in geographical scale from backyard
woodlots and roof-top solar installations to some of the largest excavations, built
structures and agricultural systems on Earth (e.g. open pit coal mines, tar sands
operations, hydro-electric dams, biofuel plantations). Regardless of their scale
or the level or type of technology deployed, all energy resource landscapes are
defined by the way they are managed principally, if not exclusively, for energy
production. An energy resource landscape implies a degree of control over space,
and access to the energy flows it can yield: oil fields, nuclear power stations and
wind farms, for example, are expressions of the social power to impose energy
production as a dominant land use, determine how and by whom energy will
be captured and transformed, and decide where energy will be distributed (i.e.
whom the resource landscape will serve). Conventional technical perspectives
place resources as the starting point of an energy system, a vital input shaping
both scale and form of economic activity and the experience of daily life
(Figure 1.1). In important ways, however, energy resource landscapes are also
outputs of the way energy systems are organised: the commissioning of hydro-
electricity plants or the development of biofuel plantations, for example, reflect
expectations about the scale and pattern of future energy use; prevailing social
norms and routines (working, travelling, leisure) that create demand for energy
services and govern the acceptability of different energy options; and the actions
(and inactions) of governments and investing companies to guide and deliver
investment in energy infrastructure. Resource landscapes are simultaneously,
therefore (i) supply zones from which societies derive their energy needs; and
(ii) spaces shaped the unequal distribution of economic and political power
within society, and which in term reproduce geographically uneven development.
This dual perspective on energy resource landscapes – as spaces of resource
provisioning and the exercise of social power – is essential to understand their
contemporary significance, and lies at the heart of this chapter.

This chapter considers the wide range of materials regarded by society as
energy resources in the context of a massive expansion of energy availability
since the Industrial Revolution. We explain the dominance of fossil fuel
extraction and thermal power generation in the global energy mix, and the
distinctiveness of the energy resource landscapes associated with fossil fuels.
We introduce the concept of resource quality, and discuss how variations in
energy density, energy return on investment, power density and carbon intensity
shape energy resource landscapes. And we show how all energy resource
landscapes (fossil or renewable, high or low carbon, sustainable or otherwise)
have environmental impacts. Throughout, the chapter highlights the contested
and political character of energy resources – i.e. how the energy resource
landscape can simultaneously be landscapes of loss, violence, and dispossession.
All resource landscapes rest on claims to property, and the authority to impose
and defend a particular form of land use. Resource landscapes are, therefore,
necessarily political: the technologies and organisational forms through which

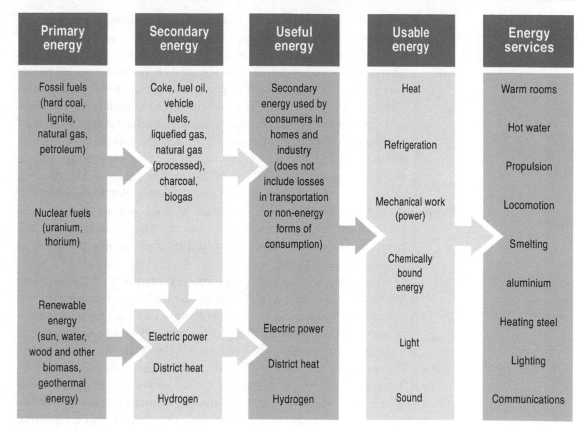

Primary energy	Secondary energy	Useful energy	Usable energy	Energy services
Fossil fuels (hard coal, lignite, natural gas, petroleum)	Coke, fuel oil, vehicle fuels, liquefied gas, natural gas (processed), charcoal, biogas	Secondary energy used by consumers in homes and industry (does not include losses in transportation or non-energy forms of consumption)	Heat Refrigeration Mechanical work (power)	Warm rooms Hot water Propulsion Locomotion
Nuclear fuels (uranium, thorium)			Chemically bound energy	Smelting aluminium
Renewable energy (sun, water, wood and other biomass, geothermal energy)	Electric power District heat Hydrogen	Electric power District heat Hydrogen	Light Sound	Heating steel Lighting Communications

Figure 1.1 Categories and conversions of energy: from energy resources to energy services (based on Bradshaw 2014 and Wagner 2009)

resource landscapes take shape, the places in which they materialise, and the communities whom they do and do not serve, are shaped as much by who is making the decisions and how decisions are made as they are by the laws of thermodynamics or supply and demand.

PORTALS OF POSSIBILITY: THE GROWING, AND UNEVEN, AVAILABILITY OF ENERGY

A rich array of materials and geophysical phenomena provide societies with the capacity to do work. Wood, animal dung, bitumen, coal, falling water, fire, municipal waste, natural gas, oats, peat, tides, timber, uranium isotopes and whale oil have, at one time or another, served as primary sources from which final energy services (heat, light, power, comfort) have been derived. These materials are, in important ways, very different things: it is not often that one finds dung, peat and uranium in the same sentence, yet they are made comparable by virtue of their potential role as **fuels**. Each is an **energy carrier**, containing

fuel
a material that releases stored chemical or nuclear energy as heat (rather than as electrical or mechanical energy). All fuels are energy carriers, but not all energy carriers are fuels: gasoline, charcoal and fissile material (e.g. uranium 235) are fuels: batteries, flywheels, and reservoirs are energy carriers but not fuels.

energy carrier

a material or phenomenon that contains energy that can then be converted to another form in order to do work. Examples include electricity, heat, logs, natural gas and hydrogen. Energy carriers enable energy to be transported and stored.

proved reserves

an estimation of the quantity of oil that geological and engineering data demonstrate can be recovered in future years, with reasonable certainty, from known reservoirs under current economic and operating conditions. The oil company BP notes that a cut-off of 90% can be used to define proved reserves: proved reserves are those that have a better than 90% chance of being produced over the life of the field.

ultimately recoverable resources (URR)

an estimate of the total amount of a resource (e.g. oil) that will ever be extracted from a field or region over time. Many resource economists either reject the idea of URR, or estimate a figure that is substantially larger than that of geologists and other physical scientists, because what can be recovered depends on technologies and market conditions and both of these are dynamic and highly uncertain over time. Global estimates of remaining ultimately recoverable resources of conventional oil vary by a factor of four between the lowest (870 billion barrels) and the highest (3,170 billion barrels).

energy that can be converted to a socially valued form (e.g. mechanical movement, chemical reaction) in a controlled manner. The reason we think of materials like coal and oil – and landscape features such as tidal estuaries and waterfalls – as potential energy resources is because they present unusually rich opportunities for capturing energy and directing it towards social goals when compared with other materials and landscapes. What are conventionally labelled 'energy resources' are, in effect, portals of possibility: they are materials or natural phenomena that offer society convenient and efficient ways to tap into the immense stocks and flows of energy that make up the physical world. These possibilities primarily concern the relative abundance of energy on offer, but they also reflect the ease with which it may be captured, stored, transported and made to flow in concentrated and controllable forms, and the acceptability and social/environmental consequences of doing so.

There is no shortage of energy from a geophysical perspective. The amount of solar energy absorbed by the Earth's surface each year is around 7,000 times greater than the energy consumed from all sources by the world's population (approximately 550 exajoules, where one exajoule is 10^{18} joules). Solar radiation is unequally distributed across the surface of the Earth with some of the highest rates in the world's hot deserts: an area of 1 million km^2 in the Sahara, for example, receives an annual input of over 8,000 exajoules, around 15 times the world's annual energy consumption (Moriartya and Honnery 2012; Candelise 2015). The total planetary stock of hydrocarbons and fissile material for nuclear power is also large, relative to consumption: **proved reserves** of oil are around 1.6 trillion barrels (or approximately fifty years of current consumption, a ratio of reserves to production that has proven remarkably stable over time), and **ultimately recoverable resources** of conventional oil may be close to twice that amount; known reserves of uranium are in the order of 5.7 million tonnes, or around ninety years at current rates of consumption. And, at a more human level, thousands of small energy conversions shape the experience of everyday life: the warmth radiating from the body of the person sitting next to you, a squeal of brakes on the morning school run, or the patter of raindrops on the pavement are all, in different ways, conversions of energy from one state to another. However, most of the planet's energy stocks and flows cannot be harnessed in meaningful or environmentally acceptable ways, and only a small fraction is available in a socially usable form.

Over time, the availability of energy has dramatically expanded (Figure 1.2) as a result of three processes. First, improvements in the technologies for capturing and harnessing energy have made it possible to extract greater energy value from long-established primary and secondary sources of energy. Examples include the invention of the wheel and gradual improvements in the collars and harnesses for draught animals, enabling a larger proportion of muscle energy to be converted into motive power; improvements in rotor blade and generator design, enabling wind turbines to convert more of the wind's mechanical energy into electricity; or the development of leaner burning internal combustion engines able to squeeze more kilometres from a litre of fuel. Second, technological development has opened up so-called unconventional energy resources, enabling society to tap some of the planet's hitherto unutilised stocks and flows of energy. The development of machinery that speeds up the rate at which energy can be applied to a task (i.e. an increase in power) has been of fundamental importance. Watt's development of the steam engine in the late eighteenth century, for example, delivered a concentrated stream of mechanical energy by transforming (via combustion) the chemical energy of coal: its effect was to open up a vast 'subterranean forest' of coal resources by enabling the pumping of water from deep mines (Sieferle 2001). Other examples of technological shifts enabling unconventional energy resource development include advances in offshore drilling and submersible pumping technology which, in the second half of the twentieth century, made possible offshore oil development; the application of nuclear fission technology to electrical power generation in the 1950s; and the development of transesterification technology for plant oils (e.g. *Jatropha)* and animal fats to produce biodiesel. Third, states and other power economic and political actors have sought to secure control over available energy resources by occupying resource landscapes claimed by others. Examples include the formation of Saudi Arabia and Iraq by Britain, France and the United States in the first half of the twentieth century, with access to the region's oil resources afforded by the Western powers' oil companies; the invasion of Kuwait by Iraq during the first Gulf War (1990–1991); an ongoing 'land grab' for plantation agriculture in the global South, including biofuel plantations (e.g. wood pellets) linked to EU renewable energy directives; or the acquisition of rural hinterlands for mines and hydropower sites by industrialising cities (White 1995).

renewable (flow) vs non-renewable (stock) energy resources
renewable energy resources, also known as flow resources, are primary energy sources with a capacity for regeneration (over timescales that are meaningful to humans) which non-renewable resources do not possess. All renewable sources are reducible to three categories (Smil 2016): solar radiation, including solar radiation's indirect and uneven effects on heating and evaporation (in the form of winds, ocean currents, and the kinetic energy of streams) and its contribution via photosynthesis to plant mass; radioactive decay in the Earth's crust, which drives the flow of heat from interior to exterior of the planet (e.g. geothermal energy); and the gravitational effect of the moon responsible for the tides.

The combination of these three processes has enabled states and societies to control a growing quantity of energy (Figure 1.2), deploying it to transform the environment and accumulate economic and social infrastructure. The amount of energy under human control has increased enormously: Grubler (2004) estimates that global energy use in 1800 stood at around 20 exajoules, increased to 50 exajoules by 1900 and by 2010 was around 500 exajoules. In short, global energy use has risen twentyfold in 200 years, while the population has increased sevenfold (Smil 2010). The result is that energy has become much more abundant over time, as measured by both the quantity available per capita and the time/energy that must be expended to acquire it. Large portions of the population have become accustomed to living in a high-energy society, while

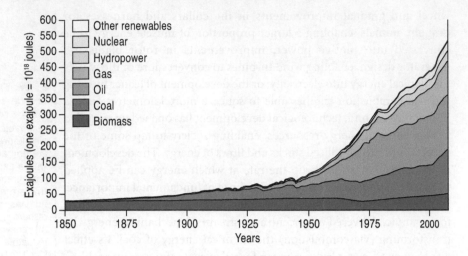

Figure 1.2 Energy availability over time (based on GEA 2013)

this is also an aspiration for most people who do not yet have it. However, the small fraction of the Earth's energy resources that can be captured, harnessed and transformed is unevenly distributed in fundamental ways. High-quality energy resources do not occur everywhere and access to resources and key technologies of energy capture and transformation is structured by prevailing distributions of wealth and power. Similarly, the social and environmental acceptability of energy resource landscapes also demonstrates social and spatial variation, as does the environmental and social consequences of energy production and transportation.

The search for socially valuable energy resources has diversified humanity's portfolio of primary and secondary energy sources over time (Figure 1.2). Simultaneously, however, a small group of primary and secondary sources have become dominant in the global energy mix. First, three hydrocarbon energy resources – oil, coal and gas – account for over 85% of all primary energy consumed worldwide. During the twentieth century, coal, oil and gas came to account for an ever-larger proportion of total energy supply. This growing role of fossil fuels in the global energy mix reflected some spatial diffusion of energy demand, but was principally a result of the enormous command over fossil energy resources enjoyed by a relatively small group of industrial countries. Towards the end of the century, however, investments in hydropower and nuclear electricity, together with the roll out of renewables, began to make a dent in the domination of fossil energy resources (Figure 1.2). Energy from these resources has doubled in the last 25 years, rising from around 10% to over 15% of the primary energy produced worldwide: hydro-electricity, nuclear power and renewables account for 6.7%, 4.4% and 2.7% respectively. Second, a growing proportion of primary energy has been converted to secondary energy sources of electricity and liquid fuels prior to final consumption. Electricity consumption has grown fourfold in the last forty years and, in that time, it has

EXAMPLE FOSSIL FUELS – WHAT'S IN A NAME?

The adjective 'fossil' derives from the original meaning of the word as 'something dug up' (from the Latin verb *fossere*, to dig). The phrase 'fossil fuels' distinguishes coal, oil and gas from other fuels, like wood or straw, gathered from the surface. The word fossil also captures the more popular notion of something old and dead, and therefore incapable of renewing itself: an account of coal in Massachusetts in 1838, for example, defines fossil fuel as "fuel which is dug out of the earth, and resulting from vegetables which have been buried there in former times" (Hitchcock 1838: 126). Coal, oil and gas are historically significant because each is a highly concentrated energy carrier (see **energy density**). Fossil fuels are a product of processes of deposition and deformation that, over millions of years, have effectively concentrated the energy content of plant and animal materials. As a result, coal, oil and gas contain more energy per unit of weight than renewable fuels gathered from the surface: a kilogram of diesel fuel, for example, has around three times the energy content of a kilogram of wood. The entry of 'fossil fuel' into the English language in the early nineteenth century corresponds, therefore, with an important shift in society's fuel gathering practices, as non-renewable, subterranean fuel sources increasingly supplemented and replaced land-based, renewable energy resources.

increased from 9% to 18% of final energy consumption (IEA 2016). As a consequence, modern civilisation depends to a remarkable degree on the appropriation of subterranean energy resources from a relatively small proportion of the Earth's surface, and their transformation and subsequent distribution as liquid fuels and electricity: in short, the experience of daily life for most people in the high-energy societies of the global North, and for a large and growing proportion of those in the emerging economies of the global South, rests on energy resource landscapes of oil and gas fields, coal mines, oil refineries and thermal power stations (see **Example**).

DIFFERENTIATED FORMS OF ENERGY: RESOURCE QUALITY VS QUANTITY

Not all energy resources are created equal. The stocks and flows of energy that provide energy services take the shape and character of particular materials (coal, wood, animal dung) and places (narrow valleys for hydroelectric dam sites, uplands for the capture and conversion of wind energy). Energy, then, is not an abstract phenomenon but embedded in material forms, geographical contexts and practices of landscape use that make energy resources much less interchangeable than inventories of energy quantity might suggest. As the late David McKay (2009), one-time chief scientific adviser to the UK's Department of Energy and Climate Change, memorably put it "You can't power a TV with

Figure 1.3 Dynamic energy resource landscapes (clockwise from top left): Bełchatów power station in Poland, Europe's largest coal-fired power station (5,400 MW) and largest single source of CO_2 emissions (photo credit: Artur Marciniec/Alamy); Solar power plant, Kanagawa Prefecture, Japan (photo credit: Σ64); Harvesting maize for biogas production in the UK (photo credit: Garnham Agriculture/Alamy); Castaic hydroelectric power plant (pumped storage), California (photo credit: Sirbatch).

cat food, nor can you feed a cat from a wind turbine." The distinction here is between different physical forms in which energy presents itself: electrical, chemical, mechanical, gravitational, for example. In short, energy resources are differentiated in socially significant ways. The notion of *resource quality* is a useful way of thinking about variations among energy resources that affect their utility and acceptability to society. The practical task of energy resource assessment, for example, is designed to disclose variations in resource quality and evaluate their commercial and social implications: wind resource assessment, for example, seeks to determine the scale of the potential wind power resource in a given area, map horizontal and vertical variation in wind power densities, identify optimal wind turbine locations, and mitigate negative environmental and social effects (e.g. impact on migratory birds). The following section identifies four significant dimensions of resource quality – energy density, power density, energy return on energy investment and carbon intensity – and how they influence the form of energy resource landscapes.

Energy density

Energy density describes the quantity of energy in a given volume or mass: it is a measure of energy concentration that can be applied to stock resources and energy carriers like wood, charcoal, peat, coal, oil and gas. Growing energy abundance over time is, to a very large degree, the result of drawing upon energy resources with progressively higher energy densities: from wood to charcoal, for example, and from charcoal and peat to coal, oil and gas. While fossil fuels are more complicated and expensive (in labour time) to capture and transform than biomass energy resources, their much higher energy densities have enabled societies to accumulate very large energy surpluses compared with organic economies. In turn, large energy surpluses have freed up the bulk of the population from energy-gathering activities and facilitated the development of complex social structures. Higher energy densities also enable societies to move and store large quantities of energy more readily, making it possible to respond to spatial and temporal variations in demand. The move towards primary and secondary energy sources with higher energy densities characterises the history of energy transitions, although not the contemporary low-carbon energy transition in which renewable energy sources play an increasing role in the energy mix.

Materials we conventionally consider as fuels have relatively high energy densities, and are also characterised by the ease (and controllability) with which they release this energy in the usable form of heat (via combustion or nuclear fission) or electricity (in a fuel cell, for example). The utility of hydrocarbons as fuels rests on the high heat values they generate when combusted (coal, oil and gas also have non-fuel values as chemical feedstocks). As their name suggests, hydrocarbon fuels are made up of molecules of carbon and hydrogen and it is the ratio of these two elements that determines their heat value, with higher H/C ratios releasing more energy when combusted. Because energy resources like methane (CH_4) have a higher ratio of hydrogen to carbon (4:1) than oil (2:1) and coal (around 1:1), the energy density of natural gas is higher than oil, which in turn is higher than coal. A kilogram of oil, for example, has nearly twice the energy content of a kilogram of coal, and around 2.5 times the same weight of wood. The practical effect of this superior concentration is that, kilo for kilo, coal and oil have a much greater capacity to do work: used as fuel in locomotives and ships, for example, oil affords a greater range than either wood or coal, while also providing similar advantages around the costs of storage, portability and transportation relative to other fuels. Set against this, however, are the larger amounts of labour that must be expended to procure a kilo of oil and coal, their more localised (i.e. less ubiquitous) occurrence, higher carbon content, and occurrence of a range of impurities (e.g. sulphur, mercury) that, depending on the circumstances of combustion, may be significant. Efforts to mitigate climate change, for example, have seen a return to wood over coal in large-scale electricity generation in parts of northern Europe: wood (and other forms of biomass) have been used as part of a co-firing of thermal power plants, such as the UK's largest at Drax (see **Case study**).

CASE STUDY: A RESILIENT RESOURCE LANDSCAPE – THE ENDURING SIGNIFICANCE OF WOOD

Wood is a primary energy source for around 2.8 billion people worldwide, and is estimated to provide almost one-tenth of the global energy supply (approximately 50 exajoules) (Bailis et al. 2015). In some forested regions of the global North, wood is a locally important fuel for heating and power generation. In the global South, fallen branches, cut timber and charcoal provide heating, cooking and an income to a much larger number of people. Charcoal is a major urban fuel in much of Africa, for example: an extensive regional trade in charcoal links forest landscapes to urban centres of consumption, while wood and other biomass (straw, dung) are important in rural areas (Kammen and Lew 2005). Although wood's significance in the global South has declined relative to other energy carriers (such as kerosene or electricity), global fuelwood harvest volumes doubled during the twentieth century (Smil 2016). In some areas of the world, including parts of South Asia and East Africa, demand for woodfuel outstrips local supply. Millions of people who rely on woodfuel must now either walk for several hours each day to gather it, or purchase wood from urban-controlled trade networks (Figure 1.4).

Wood is often regarded as a 'traditional' energy source, associated with a lack of access to modern liquid fuels or electricity. This view reflects the enduring significance of wood as a domestic fuel in the global South, but also the way Europe, North America and Japan transitioned away from wood and charcoal during the eighteenth and nineteenth centuries (Smil 2016). Charcoal provided the primary energy source for iron smelting in England until the early 1700s, when it was replaced by coke. Demand for charcoal drove significant deforestation in England, to the point that the country's growing need for iron in the sixteenth and seventeenth centuries was increasingly met by imports from Russia and Sweden. The transition to coke removed the constraint of wood supply on domestic iron output (see **organic economy**). Coke's greater energy density and power density relative to charcoal enabled iron production and consumption in England to rapidly expand.

Despite its association with traditional energy landscapes, wood is also a very modern energy resource. Forest-based biomass energy is of high importance to some countries in the global North with large forestry sectors, such as Sweden and Finland: renewable energy accounts for around a quarter of Finland's primary energy supply, with 85% of this derived from wood (Heinimo and Junginger 2009). The use of wood energy is expanding in many parts of the global North, in the context of efforts to mitigate climate change and enhance the role of renewable energy sources. Large plantations of fast-growing trees (such as willow, poplar, pine, eucalyptus and acacia) are grown to be harvested for their energy content, normally via combustion in thermal power stations where the wood is burned either on its own, or via co-firing with coal, as part of efforts to decarbonise electrical power. A rapidly growing international trade in wood pellets now links, for example, pine plantations in the southern United States to electricity companies in Europe seeking to generate renewable energy credits and meet EU targets for renewables. Alongside these 'energy forests' are two other modern forms of woodfuel, used in electricity generation or for combined heat and power: tree-waste (bark, sawdust, chips) recovered from roundwood or pulp and paper processing; and post-consumer wood recovered from the recycling of construction, packaging and other waste streams. The rise of 'energy farming' to produce solid and liquid biofuels, of which wood plantations are now a significant part, means that access to land is once again a primary determinant of energy availability (Huber and McCarthy 2017). This is placing forest lands at the centre of a struggle over how and for whom these contemporary resource landscapes should be managed.

The superior energy density of hydrocarbons has produced distinctive fossil fuel landscapes. The Industrial Revolution in Europe and North America was initially based on waterpower, but increasingly harnessed the dense energy stores of coal. By 1850 the world was getting more of its energy from coal than from wind, water and biomass sources like wood, although most coal output was concentrated in Europe and North America. The transition to coal produced entirely new resource landscapes as a result of three major effects. First, coal allowed greater quantities of energy to be obtained from a given area of land. As a consequence, energy-gathering activities became geographically concentrated around the most valuable sites. A distinctive spatial division of labour emerged, with some regions specialising in energy resource production (e.g. coal mining) and supplying a wider national (and international) economy with fuel. This process of specialisation created the classic nineteenth-century resource-landscapes associated with coal mining: densely settled mining

Figure 1.4 Women collecting firewood, Bangladesh (photo credit: Practical Action Bangladesh)

communities, a specialised mining workforce, and networks of infrastructure dedicated to extracting, processing and transporting coal. Meanwhile, coal's high energy density (compared with wood and other biomass) freed up large portions of the population from the daily task of energy gathering. Over time, most people's experience of daily life became more and more removed from the work of gathering. Older resource landscapes associated with an organic economy (wood fuel collection, crop cultivation for draught animals, or water power) became increasingly obsolete: energy resource provisioning in these landscapes had been part of a mixed, rather than specialised, spatial economy. As they did so, society's energy resource production came under the control of an ever-smaller portion of the population, giving those who had this control broader social power and making energy resources the target of corporate and state strategy.

Second, because coal, oil and gas are concentrated energy carriers, they have contributed to a growing national and international energy trade, and the increasing geographical separation of energy production and consumption. Although early water wheels were capable of concentrated mechanical power, the usable energy they generated could not be transferred far from the site of falling water. Coal introduced a new spatial flexibility to the location of industry as large quantities of energy for power could be moved far from the primary resource (coal was distinctive in this regard although not unique, see **Case study** on wood). This accelerated still further from the end of the nineteenth century, when a growing proportion of coal was used to generate electrical power. Initially, electrical generation units were market-based: power stations were built in large cities close to the demand for electrical power, with coal often transferred long distances from mines to power stations. However, with improvements in the infrastructure for electrical transmission (e.g. electrical power grids), the location of coal-fired power generation shifted back to the mines or to tide-water locations based on imported coal. The growing availability of grid electricity, powered from increasingly large, coal-fired power stations, encouraged more and more factories to shift their motive power source from coal to electricity; and, as a consequence, hundreds of small coal mines that had supplied coal to local industry ceased production. The fluid properties of oil, and increasingly gas, have pushed the geographical separation of energy production and consumption even further: two-thirds of all the crude oil extracted from the ground is internationally traded, compared with one-third for gas and around one-fifth for coal (see **Chapter 4**). In contrast with these highly stretched fossil fuel landscapes, some low-carbon energy landscapes directly re-connect the process of energy production with spaces of energy consumption. Technologies like rooftop solar, ground heat and small-scale biogas, for example, relocate energy production to the site of consumption.

Third, the rate and rhythm of energy capture via fossil fuel resource landscapes are unlike those associated with biomass, wind and water. The rate of energy capture and conversion by mines and power stations is much greater than that of the **organic economy**, based on muscle power and supplemented by wind and water. There are also important qualitative differences, as the output of energy from fossil fuel landscapes is not set by the solar flux, seasonal water flow, or the rate of regeneration of biomass. As coal use expanded in the eighteenth and nineteenth centuries, constraints on energy production increasingly lay in the scale of technology and capacity to control labour. Freed from the 'photosynthetic constraint' of the organic economy, the rhythms of production became organised around market or state logics (rather than daily or seasonal variations in sunlight or water availability). Such logics have driven a search for scale economies in energy production (bigger mines, larger power stations) which, in turn, required scale economies in energy transportation and storage (larger coal trains, crude oil carriers, and oil and gas storage facilities). New techniques of energy capture, such as open pit mining and processes like mountain top removal, have facilitated growing economies of scale. Today, the largest coal mines in the world, in the Powder River Basin of Wyoming, are capable of producing around 100 million tonnes per year, while a number of mines in Indonesia and China have a capacity of around 50 million tonnes.

organic economy
a term introduced by the economic historian, E. A. Wrigley (1988), as part of his account of the Industrial Revolution. It describes an economy in which solar energy flows, fixed via vegetation, are the dominant source of energy and raw materials. In an organic economy, the productivity of the land surface (i.e. its ability to capture and transform solar energy flows) represents the primary constraint on economic growth. Wrigley contrasts this with the mineral-based energy economy: here societies supplement solar flows by accessing accumulated stocks of energy (e.g. coal) and so the productivity of land is no longer a constraint on growth.

The rise of 'extreme energy': energy return on investment (EROI)

Capturing and harnessing energy resources can take a good deal of work. Energy resource landscapes are characterised by the application of energy (in construction, digging, pumping and processing, for example) as well as its capture and transformation. The balance between these input energy costs and energy outputs describes a resource landscape's net yield of energy, or what is often referred to the 'energy return on energy invested' or, more simply, energy return on investment (EROI). Most energy resource landscapes provide society with a sizeable return on the energy expended in capturing, harnessing and transforming energy resources and this 'surplus' is why they are regarded as energy resources in the first place. However, EROI provides a way to compare among different energy resource landscapes, as well as to chart significant changes in the net energy yield of particular landscapes over time.

Year-on-year increases in energy availability (Figure 1.2) are a striking feature of the environmental and energy history of the twentieth century. The growing quantity of fossil energy resources that have underpinned this growth are all the more remarkable, given that generally they have been secured from progressively lower quality resource stocks. Large reserves of low-cost, easy-to-access oil remain in the Middle East, but much of the 'easy oil' elsewhere has been exploited. As a result, the energy return on energy invested has been declining for most fields around the world. In the US, for example, the average EROI for finding and producing domestic oil declined from 100:1 in the 1930s to about 30:1 in the 1970s, and to around 15 today. Worldwide, the average EROI for oil and gas is estimated to have risen from 26 to 35 in the late 1990s – as low oil prices allowed oil producers to concentrate their efforts on higher quality oil and gas resources – but it subsequently fell to around 19:1 (Hall et al. 2008: 117). An increasing proportion of future oil production will come from deepwater offshore fields, and from unconventional oil resources such as shale or bituminous sands. In general, these energy resources have lower EROIs than the conventional crude oil that has under-pinned decades of growth in oil demand because of the greater energy inputs required for extraction and processing. New developments are moving to more complex settings such as deeper wells (higher temperatures and pressures) and offshore environments so that costs rise. These shifts to more challenging and costly environments suggest, to some analysts, the end of 'easy oil' and rise of 'extreme energy' (Klare 2010). Development of these high-cost oil resources will require future oil prices to be sustained at levels that are two to four times above historic price norms of around $25 per barrel: it is not at all clear that future levels of demand will be sufficient to generate these prices. In the same way, operating costs in mature oil and gas fields tend to rise over time, even though the initial costs of infrastructure may have been borne long ago. The *resource pyramid* (Figure 1.5) captures the relationship between an apex of high-quality resources that are cheap to extract in energy and economic terms, and a larger mass of more diffuse, lower-quality materials (Ahlbrandt and McCabe 2002). Efforts to secure oil by moving down the resource pyramid and tapping lower quality, non-conventional fossil fuels such as bituminous sands – or conventional crude oil in unconventional locations like the deepwater offshore or Arctic – reduces EROI ratios further: estimates for bituminous sands, for example, range from 10:1 to less than 1:1 (implying a net loss of energy). Falling EROI ratios for oil also explain why oil *producers* are now among the largest corporate emitters of carbon dioxide: large quantities of fossil energy are consumed in the process of extracting crude oil and/or unconventional fossil fuels and processing them into secondary energy resources (gasoline, diesel, fuel oil etc).

Over time, the movement down the grade curve/up the cost curve implies the opening of new resource landscapes as the centre of gravity of energy gathering activities shifts towards lower-quality resources. However, the evolution of resource landscapes does not always follow the path assumed by the resource pyramid. Disruptive technological changes – such as horizontal drilling and hydraulic fracturing for shale oil and gas resources – can also open new frontiers, as can regulatory decisions that absorb the costs/risk of new development. Geopolitical considerations can prolong the life of existing resource landscapes beyond what the 'market' may determine. The desire of national governments for energy supply security, or considerations about (un)employment, mean states may favour indigenous low-grade primary energy sources rather than import higher grade alternatives (**Chapter 4**). The history of coal mining in Europe, and the continuing dominance of domestic coal in the energy mix of Germany and Poland, illustrates this dynamic: coal continues

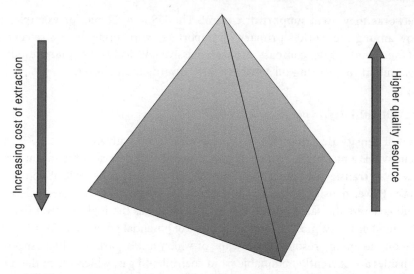

Figure 1.5 The resource pyramid (based on Ahlbrandt and McCabe 2002)

to supply 85% of electricity in Poland and around 50% in Germany notwith-standing the EU's decarbonisation agenda. Much of this coal comes from low-grade lignite mines within Poland and Germany, a domestic energy resource that is favoured amid concern about Russia's dominance in European gas markets. Furthermore, societies do not possess complete geographical knowledge, and so in practice lower-quality resources have often been used before high quality ones were discovered or became available. This phenomenon is described by the ecological economist, Richard Norgaard (1990), as the 'Mayflower Problem' in reference to European settlers (the 'Puritans') who landed on the eastern seaboard of North America in 1620 aboard the *Mayflower*. Norgaard (1990: 23) points out how

> the history of North America . . . is a history of using low quality resources before learning led to the exploitation of higher quality, less costly resources. Many generations passed before American agriculture shifted from the relatively poor soils of the east coast to the more productive Midwest.

Tapping and controlling high-quality sources of supply, then, is one strategy for handling variations in the quality of energy resources. Here the Mayflower analogy is particularly appropriate because, as the bloody history of European settlement in North America illustrates, one person's discovery is another's dispossession: the capacity to claim and harness energy resources is not solely a function of knowledge but also political (and ultimately military) control. Annexing new energy supplies or controlling energy flows beyond a nation's borders can be a means to keep domestic prices for energy low in the face of localised depletion. Regional and global powers have the capacity to produce highly stretched resource geographies via production and transportation networks they own, support or control. The US and China, for example, are now among the world's primary oil importers, with supply lines stretched to all four points of the compass and with massive political and military resources committed to ensuring oil's orderly flow towards their shores.

Power density: spatial trade-offs

Power density describes the flow of power from a given area. Like energy density it is a measure of concentration, but one that describes the rate at which energy is transferred rather than the content of energy per unit of volume or mass. Power density is measured in Watts/m^2. If two unequally sized pieces of land generate the same power, the smaller piece has the higher power density. Put another way, a piece of land has a range of potential power densities depending on the energy resource landscape of which it is a part. The highest power densities are currently obtainable from coal, oil and gas, which are in the range of 400 to 4,000 W/m^2 (Smil 2016). Renewable energy landscapes have power densities several orders of magnitude lower than resource landscapes associated with fossil fuel extraction and thermal power generation. To harness an equivalent flow of power from renewable energy sources requires a much larger area of land be dedicated to the task than for fossil or nuclear energy sources.

Historically, land scarcity has limited how much power biomass-based societies are able to harness: the scale of iron production in England was constrained by the availability of forest land as early as the sixteenth century, as it relied on charcoal (derived from wood) prior to the development of iron-smelting techniques using coke (derived from coal). Although both charcoal and coke are almost pure carbon and have a similar energy content (i.e. their energy densities are about the same), coke has a power density around 7,000 times higher than charcoal. The application of coke to iron production enabled output of iron (and subsequently steel) to surge in Europe and the United States during the nineteenth century. Smil (2016) reports that the US iron production expanded 500-fold between 1810 and 1910, by which time the annual output of 25 million tonnes would have required an area of forest the size of Wisconsin for fuel in the absence of the switch to coke (see **Case study**).

The lower power density of renewable energy resources has three significant implications for resource landscapes. First, the logistics of energy capture and distribution are different to those for conventional fossil fuel extraction and thermal power generation. As outlined above, fossil fuel landscapes are characterised by the control of highly concentrated energy sources, the transformation of energy into usable forms (power stations, oil refineries) at very high power densities and in a relatively small number of locations, and the subsequent distribution of energy to end users. By contrast, harnessing flow resources like wind or solar for electricity production requires co-ordinating multiple dispersed locations. This is necessary to overcome the relatively low power densities of these sources, and to manage intermittency in supply (e.g. daily or seasonal variation in levels of sunlight, or the reliability of wind).

The implication of harnessing renewables, therefore, is a growth in the number of energy resource landscapes, their potential overlay upon (and interaction with) other land uses, and the development of new network forms in which non-contiguous landscape elements are nonetheless managed as a whole. Second, the higher power densities of fossil fuel extraction and thermal power generation means that a given level of investment in control/exclusion of other potential users provides command over a much greater power flux. Control of key sites (oil fields, major power stations) therefore affords tremendous political leverage. The lower power density of renewables makes it harder to replicate this level of control from single sources and, potentially, increases the security of supply. Conversely, however, the profusion of suppliers creates new challenges of co-ordination and infrastructure connectivity (see **Chapter 3**).

Third, the lower power density of renewable energy resources means that energy capture is once again a significant driver of land use and vegetation cover (Howard et al. 2009; Huber and McCarthy 2017). Decarbonisation and sustainable energy goals target a greater use of renewables, including the 'return' to a bio-economy of energy production based on the cultivation and harvesting of biomass and biofuels, and the collection and treatment of food and agricultural wastes. In this respect, animal dung, straw and wood have gone full circle (see **Case study**), although the challenges of a modern bioeconomy of energy capture are to do this efficiently, at scale and in ways that are sustainable (Scarlat et al. 2015). Biomass and biofuels include quick-growing woody crops like willow and poplar, and grasses such as *Miscanthus sinensis* (Elephant Grass) for co-firing in electricity generation or the production of low-carbon heat; the cultivation of starch (wheat, maize) and sugar crops (sugar cane, sugar beet) for bioethanol production; and oil crops (rapeseed, soybean, jatropha) for biodiesel production. Energy from waste includes biogas (renewable natural gas) via anaerobic digestion of food and agricultural waste and waste water, and recovery of gas from landfills. Because land requirements of biomass and biofuels are significant, recent initiatives to expand biomass production in the interests of decarbonisation and energy security have brought it into confrontation with both food production and biodiversity conservation targets. Growing concerns that biomass and biofuel production compete for land with food supply and habitat production, rather than utilising agricultural or forestry wastes, has led to some recalibration of policy. The EU, for example, has a target of 10% of transport fuel from renewable sources by 2020 but amended this in 2015 to cap the amount supplied from crops grown on agricultural land to 7% (Blaschke et al. 2013; Scarlat et al. 2015).

Carbon-intensity: from high to low carbon resource landscapes

Concern about global climate change caused by the accumulation of greenhouse gases (GHG) in the atmosphere is already shaping the formation of energy resource landscapes in distinctive ways. The energy system accounts for around

65% of global GHG emissions and is widely acknowledged as the most important source of carbon dioxide. Other important contributions to climate change from the energy sector include fugitive methane emissions from natural gas production plants and pipelines, flaring of methane associated with oil development (e.g. around shale oil in North Dakota), and hydrofluorocarbons used in air-conditioning applications (for which a landmark international phase out was agreed in Rwanda in 2016). Climate change is not the only environmental problem related to energy production and consumption, or even the most significant. Urban air pollution linked to power generation and transport, and poor indoor air quality associated with the use of traditional fuels for cooking and heating, are now major drivers of energy system change in many parts of the world (see **Chapter 7**). Climate change, however, is unparalleled in its global causes and consequences, and in the extent to which efforts to address it will require major changes to energy resource landscapes around the world.

Scientific consensus is that global climate change introduces a fundamental 'carbon-constraint' on the way in which energy systems (and economies more generally) must develop. The United Nations Framework Convention on Climate Change, adopted in Paris in December 2015, confirmed that global carbon emissions must be reduced by 80% by 2050 if the world is to avoid a rise in temperature of more than 2°C. The Paris Agreement also acknowledged that emissions in some places may need to rise, to ensure access to high-quality energy services. Embedded in regulation and policy at urban, national and international levels, the effect of the carbon constraint is to differentiate the energy sector by reference to its carbon intensity. A goal of climate policy is to 'decarbonise' energy use by promoting low carbon energy solutions and, to a lesser extent, actively phasing out high carbon forms of energy, as well as investing in energy efficiency and demand reduction. The energy sector's major contribution to greenhouse gas emissions means it will carry much of the weight of efforts to decarbonise national economies. Electricity generation is a favoured policy target, although transportation and heating (and cooling) are also common objectives of decarbonisation policies. As we outline below, several distinctive resource landscapes are emerging associated with patterns of investment and financial flows enabled by climate policies. At the same time as new landscapes are emerging, established resource landscapes associated with the 'old carbon economy' of fossil fuel extraction and combustion are showing signs of contraction in some parts of the global North. There are also signs that investment is being deflected away from high-carbon to low-carbon energy sources in emerging economies. The effect of the carbon constraint, therefore, is geographically very uneven and increasingly dynamic: while coal mining and the use of coal in electricity generation have sharply contracted in the US and UK in the last few years, for example, new coal resource landscapes have been commissioned in Germany, South Africa, Turkey, Australia, China and India. We illustrate below how measures of relative carbon intensity are already having significant effects on resource landscapes. Honouring the Paris Agreement, however, will require radical changes to resource landscapes, including the rapid retirement of large chunks of infrastructure and resources currently dedicated to capturing and transforming fossil fuels; while, at the same time, creating new landscapes associated with low carbon energy systems.

Schematically, three different types of resource landscape changes are associated with the emergence of a carbon constraint. *First*, the quest for low-carbon solutions is 're-basing' the energy sector around a different set of natural resources, enrolling materials and landscapes into the energy system that in modern times have previously lain outside it. From the growth of biomass plantations and offshore wind farms to the installation urban solar panels and anaerobic digesters, the proliferation of low-carbon resource landscapes is among the most visible manifestations of how existing energy systems are being re-configured by the carbon-constraint. Many of these landscapes often materialise in (rural) places not previously utilised as sites of energy capture, and which are often valued for their production of other environmental goods and services. Their scale, and the novelty of the energy resource land use, mean they are often experienced as intrusion and imposition, and sometimes as exclusion with the loss of land access. Low-carbon energy resources are inextricably bound up with the politics of land (McEwan 2017; Yenneti et al. 2016). In this regard, they are no different to conventional resource development or the shift towards unconventional oil and gas: implementation without prior informed consent from affected communities and their participatory involvement in decision-making constitutes a 'colonial' form of resource control.

Climate change policies, and the availability of climate finance for projects that can help mitigate climate change, are steering a growing proportion of energy investment into the capture, harnessing and transformation of low-carbon energy resources. The dominance of hydrocarbons in the global energy mix means that the levels of investment in low-carbon energy capture are still small by comparison, but the rate of capacity expansion is impressive. Japan, for example, experienced a rapid solar boom in the five years following the Fukushima nuclear meltdown as the government incentivised low-carbon power alternatives, including offering a feed-in tariff for new solar. Capacity increased sevenfold between 2011 and 2015 (notwithstanding tariff cuts) making Japan the second largest PV market (after China) and the third largest installed PV capacity (after China and Germany). Initially led by large, utility-scale solar farms (including installations on abandoned golf courses constructed at the height of the property boom in the 1990s and early 2000s), the market has shifted to residential roof-top installations, which are expected to account for half of the annual solar market by 2020 (Bloomberg Technology 2016; *Independent* 2015). Climate finance is a significant driver of low-carbon energy landscapes in the global South. The UNFCCC's Clean Development Mechanism (CDM) – initiated as part of the Kyoto Protocol in 1992 – for example, enabled companies in the global North to offset their emissions by purchasing emission reduction credits from projects in the global South. Through the CDM and other mechanisms, climate finance has spurred the creation of wind farms, afforestation schemes, off-grid solar power, programmes to replace household cooking stoves, and initiatives to capture methane from landfill in developing economies. While many of these new landscapes capture, harness and transform energy resources, their function is primarily to create emission reduction credits that can be purchased by companies operating in the global North. They operate as 'exclaves' or geographical annexes to fossil fuel resource landscapes in the global North, with each landscape co-producing the other. Over 8,000 CDM projects have been established, but the effectiveness of the programme in reducing emissions has been widely questioned and demand for emission reduction credits has largely dried up. A similar programme, learning from the challenges of the CDM, is taking shape under Article 6 of the Paris Agreement.

While solar heat and power, and a return to wind and bioenergy (discussed above, see also **Case study**) are among the most striking low-carbon energy trajectories, the search for non-fossil energy carriers has also included a re-evaluation of the potential for nuclear energy, and initiatives to develop a hydrogen-based energy economy involving a growing use of fuel cells. The new materials of a low-carbon energy economy are not only fuels or energy carriers, however: they also include lithium and cobalt for battery technology, and a suite of rare earth elements which, despite their name, are relatively abundant in total quantity but occur in low concentrations so that their extraction and processing are difficult and costly, with significant environmental consequences. Neodymium, for example, is a key component of the permanent magnets used in direct-drive wind-turbine generators able to operate at low-wind speeds; and indium and tellurium are used in thin film technology for photovoltaics. The production of rare earth elements has doubled in the past twenty years with over 90% of production coming from China. Mines and processing plants near Baotou, in the Inner Mongolian region of northern China, for example, supply around 40% of the world's demand and have transformed large areas of pastureland into open pits, waste dumps and tailings ponds.

Second, within the 'old carbon economy' of fossil fuels, the relative carbon intensity of different hydrocarbons is assuming a significance it did not have prior to concerns about climate change. Intensification of this trend will result in a shakeout of higher-carbon fossil fuels (particularly coal) and a flight of investment towards lower-carbon fossil fuel options such as natural gas, although currently the pattern is very mixed. There are important trade-offs with energy security and affordability, however, so countries with large domestic coal reserves (such as India and South Africa) may stick with coal for some time. The CO_2 emission potential of hydrocarbons is a function of the ratio of hydrogen to carbon: hydrocarbon fuels with a high proportion of hydrogen atoms to carbon have a lower CO_2 emission potential per unit of heat, so that methane (CH_4, with a ratio of 4:1) produces less CO_2 when burnt than either oil or coal for the equivalent amount of heat. This differential carbon intensity is recognised in policy initiatives that promote lower-carbon fuels and in market mechanisms (like the EU Emissions Trading System (EUETS), the Regional Greenhouse Gas Initiative of nine northeastern states in the US, and China's national ETS planned for 2017), which put a price on carbon, effectively making higher carbon fuels more expensive for power stations and industrial plants to consume. In the electrical power generating sector, the effect of this differential carbon intensity is to discriminate against coal in favour of natural gas.

At the moment, the price of carbon is too low for carbon intensity alone to drive a rapid transition away from high carbon fuels. A shift away from coal in power generation is occurring in many countries of the global North, but it is the combination of cheaper fuel costs, quicker construction times and public health regulation (e.g. the EU's Large Combustion Plant Directive, and Mercury and Air Toxics Standards in the United States) that is driving the transition to gas rather than carbon intensity. Policies targeting coal-fired power generation, therefore, can deliver a series of 'co-benefits' besides mitigation of climate change. Nonetheless, the potential 'savings' in emissions available by switching from coal to gas in power generation (around 50% at the point of combustion) mean natural gas aligns much more readily with the direction of travel of climate policy, enabling proponents of natural gas to position it as a 'bridging fuel' facilitating the transition to a lower-carbon energy future. The status of gas as a bridging fuel, however, critically depends on it replacing existing higher-carbon generating capacity, and pricing structures that will drive the replacement of natural gas over time with non-fossil alternatives. Without these, growing use of natural gas effectively locks-in a carbon-intensive energy system and blocks the transition needed to low-carbon sources.

The lower greenhouse gas emissions profile of natural gas relative to coal or oil has contributed to a growing demand for natural gas which, in turn, has propelled a growth in gas extraction and the infrastructures for gas processing, storage and transport. Worldwide gas consumption has grown 25% in the last decade, and by 2040 gas is expected to overtake coal to become the second most important fossil fuel after oil. The result is a series of new resource landscapes dedicated to mobilising natural gas (Figure 1.6) which include: the fracking of 'unconventional' shale gas resources in the United States; a proliferation of infrastructure (pipelines, liquefaction and regasification terminals, shipping) for gas transportation; sharp increases in both conventional and unconventional (e.g. coal seam) gas production, and associated massive investment in export infrastructure, in Australia; and an expanding use of gas for electricity generation. In the US, for example, expansion of electricity generating capacity via gas-fired power plants has run well-ahead of all other energy sources for several years. Fed by growing domestic supply from the fracking of shale resources, electricity generation via gas has caused sharp declines in coal-fired generation (from around 60% of electricity to just over 30% in less than a decade) and has hit the coal mining sector hard: the contribution of natural gas to US electrical supply surpassed coal for the first time in 2015. In **Chapter 4** we discuss how international climate commitments and policies to promote decarbonisation have the potential to 'strand' high-carbon resources and infrastructures, such as coal-fired power stations and reserves of coal and oil reserves.

Third, low-carbon resource landscapes also include the emergence of technologies, infrastructures and practices that seek to capture carbon dioxide

Figure 1.6 Gas resource landscapes (clockwise from top): Ocean-going tanker carrying liquefied natural gas (photo credit: JoachimKohlerBremen); Public bus powered by compressed natural gas, Nebraska (photo credit: Hanyou23); Laying a gas distribution system in an informal settlement on the outskirts of Istanbul, Turkey (photo credit: Elvan Arik); Installing domestic gas supply in an apartment building in Sfax, Tunisia (photo credit: Hugo Bolzon).

at the point of combustion and before it is released to the atmosphere. In a carbon-constrained world, carbon capture and storage (CCS) becomes as essential to the continued use of fossil fuels as was the atmosphere's availability as a free 'dump' for CO_2 during the last three hundred years. There are both technical and economic challenges to the implementation of CCS and, at the current time, its application is limited to a small number of sites worldwide. CCS is not an option for distributed sources of CO_2 emissions (such as transportation or housing), but it has potential for large, point-source emissions such as coal or gas fired power stations. However, at the moment the market price of carbon is too low to drive the uptake of CCS. Delivering CCS at scale will require both extensive collection and transmission networks (i.e. pipelines) for transporting CO_2, and the availability of a low-cost and reliable subterranean repository. Abandoned oil and gas fields represent potential storage sites, so that areas like the North Sea – relatively close to large energy markets in Europe, with extensive pipeline infrastructure in place, and where geological conditions are relatively well known – could be re-appraised for their value as carbon dioxide stores rather than supply zones for oil and gas.

CONCLUSION: RESOURCE LANDSCAPES OF SOCIAL POWER

In this chapter, we have introduced a diverse set of energy resource landscapes and described some of their primary characteristics. Our goal has been to move beyond energy as an abstract concept by examining some of the material characteristics of resource landscapes. We have identified important differences among energy resources and shown how these create differential opportunities and limits for social use, and highlighted the social and geographical unevenness of resource access. The resource landscape perspective is useful because it restores the particularity, distinctiveness and incommensurability of different energy resources – characteristics hidden from view by the technical concept of energy, invented in the nineteenth century.

The availability of energy has increased far faster than population growth, enabled primarily by the shift to fossil resources with higher energy densities and, subsequently, by the rapid and sustained expansion of output of coal, oil and gas. Just as significant has been the development of machines fuelled by fossil energy resources capable of delivering very high power densities. These have enabled intensive and far-reaching environmental transformation, locking in demand for resource landscapes that are able to achieve these power densities. As a consequence, a growing proportion of final energy supply has come in the form of liquid fuels and gas rather than solid fuels, and via forms of delivery that are grid connected (e.g. gas and electricity distribution). Worldwide, around a third of the final energy reaching consumers currently does so in solid form (e.g. wood or coal), another third as liquids (primarily for transportation), and a third via gas and electricity distribution grids (IPCC 2007). Over time, the shift to more energy dense resources has freed up people from the task of energy gathering, enabling specialised energy resource landscapes to emerge and a growing geographical separation of energy resource gathering activities from spaces of consumption. This has been accompanied by a process of abandonment at the local scale, as lower-quality resource landscapes (small coal mines, marginal oil fields, harvesting of peat and wood) cease to function. As a consequence, the experience of 'modernising' a community's energy supply frequently involves a growing reliance on higher quality energy resources collected and processed elsewhere. At the same time, the increasing significance of long distance resource supply systems has enhanced competition for control the highest quality resource landscapes, and the challenges of supply security and territorial control (see **Chapter 4**).

The drivers of new resource landscapes are diverse, from national energy security concerns prioritising development of domestic energy resources and opportunities for transnational capital to achieve returns on investment, to modernisation of energy supply and incentivising economic development. The recognition of a global carbon constraint introduces a new logic which is now shaping resource landscapes in significant ways, as part of a low-carbon energy transition (see **Chapter 10**). Importantly, however, reaction to the carbon constraint and the dilemma of developing secure, affordable and equitable resource landscapes that are environmentally benign has taken a variety of paths. Low carbon energy landscapes have rapidly proliferated over the past decade, in both the global North and South. Progress on disassembling fossil intensive resource landscapes is much patchier, however: there is some movement away from coal-fired power generation, although new coal resource landscapes (to say nothing of oil and gas) continue to be built into development trajectories in many parts of the world. The economic forms these and other energy landscapes take are explored in **Chapter 2**.

The everyday sense we have of resource landscapes such as oil fields, power stations or coal mines tends to conceal the conditions of their possibility and emergence. Capturing and harnessing energy is an act of appropriation, a process of taking (in the form of property) which depends on systems of legal and extra-legal support through which property claims can be secured, realised and defended. These include, for example, the state's powers of eminent domain to acquire private property for public use. Energy resource landscapes, therefore, are always in a state of 'becoming', in that their existence as resource landscapes is contingent on their relation to wider social, economic and political processes. Even where the sense of a resource landscape may be widely shared, it is never a *universal* view: conflicts over the construction of large dams or wind farms, for example, readily expose the range of value systems at play around natural resources and demonstrate how one group's energy resource can be another's dispossession. Resource landscapes are frequently contested spaces, where the status of 'resource' is challenged by a range of alternative discourses, including notions of rurality, wilderness, indigeneity, social justice, and alternative horizons of economic value. Energy resource landscapes not only capture, harness and transform energy: as we explore in **Chapters 2 and 3**, they also (re-)produce economic and political power.

QUESTIONS FOR DISCUSSION

- Explore the energy flows associated with normal daily routines; what sorts of resources provide these energy inputs? Can you find out where any of these resources come from? Why is it difficult to trace some of the connections?

- Identify several locations where electricity is generated in your area (wind farm, roof-top solar installations, thermal power plant) and consider differences/similarities in the kind and scale of energy conversions: in what ways are these related to the material characteristic of the resources? Using publicly available data, place these specific installations into a broader perspective.

- Consider an important energy service like (domestic) heating: how have the resource landscapes associated with this service changed over time?

- Argue for/against the proposition that shale gas/tar sands/upland wind is an energy resource. Reflect on *for whom* these may be resources.

ACTIVITIES/POTENTIAL RESEARCH PROJECTS

- Identify a contemporary energy resource development controversy: conduct research to work out who decides whether this project goes ahead, and how this decision is made. Consider the range of arguments put forward for and against the development: who (socially, geographically) are imagined to be the winners and losers?

- How can concepts like 'enclosure' and 'dispossession' be used to understand conflicts over 'green' power schemes like large-scale hydro-electric power plants, wind farms and biomass plantations?

RECOMMENDED READING

■ Baka, J. 2016. Making space for energy: wasteland development, enclosures, and energy dispossessions. *Antipode* 49 (4): 977–996.

Drawing on fieldwork in southern India, this paper explores the social consequences of policies designed to stimulate biodiesel production. It shows how biodiesel plantations, based on the field crop *Jatropha*, took shape in rural areas in response to urban demands for liquid fuels able to supply modern energy services. To make room for these plantations, however, land had first to be cleared of a tree species (*Prosopis*) that was providing fuelwood for households and industry. Because Jatropha and Prosopis serve different social groups – Jatropha was used to produce liquid fuels that supplied urban rather than local energy needs – the author argues that biodiesel production was experienced by local communities as a process of 'energy dispossession'.

■ Evenden, M., 2015. *Allied power: mobilising hydro-electricity during Canada's Second World War*. Toronto: University of Toronto Press.

China, Brazil and Canada are hydroelectricity super-powers, as measured by total electricity output. *Allied Power* tells the story of how Canada attained this status, focusing on the efforts of business and government interests to harness the power of Canadian rivers during World War I. Canada's hydroelectricity generating capacity soared as a result of a national drive for cheap power to feed aluminium smelters and other wartime industries. This fascinating account of the making of an extensive and variegated energy resource landscape examines the political and social forces that transformed Canadian rivers in the middle of the twentieth century, and the lasting consequences of an energy planning process that 'spoke of plants and power, not people and place'.

■ Lahiri-Dutt, K. 2016. The diverse worlds of coal in India: Energising the nation, energising livelihoods. *Energy Policy* 99: 203–213.

Coal dominates energy production in India. The country's policymakers and planners tend to think of coal's resource landscapes in national terms, focusing on how the country's mines and power stations deliver energy security, development and prosperity. However, Lahiri-Dutt shows how in practice there are five overlapping coal economies. These are differentiated by ownership, labour regimes and the norms and values that govern production. The author explores the linkages among these multiple and diverse worlds of coal, and shows how coal mining is embedded in livelihood practices in ways that complicate narratives of national development or energy transition.

■ Smil, V. 2016. *Power density: a key to understanding energy sources and uses.* Cambridge, MA and London: MIT Press.

The distinguished energy geographer Vaclav Smil, Emeritus Professor at the University of Manitoba, has made numerous contributions to understanding the evolution and distinctiveness of the contemporary energy system, and the socio-economic significance of the physics of energy conversions. This very readable book explores the importance of power densities (i.e. how much power is produced, or used, per unit area of the Earth's surface). The question of the land area necessary to sustain a given power output is increasingly important in the context of efforts to scale up the use of renewable energy resources, which have much lower power densities than fossil fuels. The concept at the heart of this little book provides a way of thinking through some of the spatial trade-offs involved in the transition towards low-carbon energy sources.

REFERENCES

Ahlbrandt, T. and P. McCabe. 2002. Global petroleum resources: a view to the future. *Geotimes* 47(11): 14–18.

Bailis, R., R. Drigo, A. Ghilardi and O. Masera. 2015. The carbon footprint of traditional woodfuels. *Nature Climate Change* 5: 266–272.

Blaschke, T., M. Biberacher, S. Gadocha and I. Schardinger. 2013. 'Energy landscapes': Meeting energy demands and human aspirations. *Biomass and Bioenergy* 55: 3–16.

Bloomberg Technology. 2016. Japan's solar boom showing signs of deflating as subsidies wane. 6 July 2016. Available online at www.bloomberg.com/news/articles/2016-07-06/japan-s-solar-boom-showing-signs-of-deflating-as-subsidies-wane

Bradshaw, M.J., 2014. *Global energy dilemmas: energy security, globalization, and climate change.* Polity Press.

Candelise, C. 2015. Solar energy: an untapped growing potential? In *Global energy: issues, potentials and policy implications.* P. Ekins, M. Bradshaw and J. Watson. Oxford: Oxford University Press, pp. 354–376.

GEA, 2012: *Global Energy Assessment - Toward a Sustainable Future,* Cambridge University Press, Cambridge, UK and New York, NY, USA and the International Institute for Applied Systems Analysis, Laxenburg, Austria.

Grubler, A. 2004. Transitions in energy use. In *Encyclopedia of Energy.* C. J. Cleveland (ed.), Vol.6, Amsterdam: Elsevier, pp. 163–177.

Hall, C.A., R. Powers and W. Schoenberg. 2008. Peak oil, EROI, investments and the economy in an uncertain future. In *Biofuels, solar and wind as renewable energy systems.* The Netherlands: Springer, pp. 109–132.

Heinimo, J. and M. Junginger. 2009. Production and trading of biomass for energy – an overview of the global status. *Biomass and Bioenergy* 33: 1310–1320.

Hitchcock , E. 1838. *Report on a re-examination of the economical geology of Massachusetts.* Amherst, MA: Adams Printers.

Howard, D., Richard A. Wadsworth, Jeanette W. Whitaker, Nick Hughes, Robert G.H. Bunce. 2009. The impact of sustainable energy production on land use in Britain through to 2050. *Land Use Policy* 26S: S284–S292.

Huber, M.T. and McCarthy, J., 2017. Beyond the subterranean energy regime? Fuel, land use and the production of space. *Transactions of the Institute of British Geographers.* International Energy Agency 2016. *World Energy Outlook.* Paris.

The Independent. 2015. Japan is turning its abandoned golf courses into solar power plants. Monday 20 July 2015. Available online at www.independent.co.uk/sport/golf/japan-is-turning-its-abandoned-golf-courses-into-solar-power-plants-10402042.html

IPCC. 2007. IPCC fourth assessment report: climate change. Working Group III: mitigation of climate change, section 4.3.4 energy carriers. Available online at www.ipcc.ch/publications_and_data/ar4/wg3/en/ch4s4-3-4.html

Kammen, D. and D. Lew. 2005. *Review of technologies for the production and use of charcoal.* United Nations.

Klare, M. 2010. The relentless pursuit of extreme energy. Washington, DC: Institute of Policy Studies. Available online at www.ips-dc.org/the_relentless_pursuit_of_extreme_energy/

McEwan, C. 2017. Spatial processes and politics of renewable energy transition: land, zones and frictions in South Africa. *Political Geography* 51: 1–17.

McKay, D. 2009. *Sustainable energy without the hot air.* Available online at www.withouthotair.com

Moriartya, P. and D. Honnery. 2012. What is the global potential for renewable energy? *Renewable and Sustainable Energy Reviews* 16: 244–252.

Norgaard, R.B. 1990. Economic indicators of resource scarcity: a critical essay. *Journal of Environmental Economics and Management* 19(1): 19–25.

Scarlat, N., and J.-F. Dallemand, F. Monforti-Ferrario and V. Nita. 2015. The role of biomass and bioenergy in a future bioeconomy: policies and facts. *Environmental Development* 15: 3–34.

Sieferle, R. 2001. *The subterranean forest: energy systems and the industrial revolution.* Cambridge: White Horse Press.

Smil, V. 2010. *Energy transitions: history, requirements, prospects.* Denver, CO: Praeger.

Smil, V. 2016. *Power density: a key to understanding energy sources and uses.* Cambridge, MA and London: MIT Press.

Wagner, H.-J. 2009. *Energy: the world's race for resources in the 21st century.* London: Haus Publishing.

White, R. 1995. *The organic machine: the remaking of the Columbia River.* New York: Hill and Wang.

Wrigley, E.A., 1988. *Continuity, chance and change: The character of the industrial revolution in England.* Cambridge: Cambridge University Press.

Yenneti, K., R. Day and O. Golubchikov. 2016. Spatial justice and the land politics of renewables: dispossessing vulnerable communities through solar energy mega-projects. *Geoforum* 76: 90–99.

Economic landscapes

Development driver. Paycheque provider. Public good. Profit centre. Economic landscapes are shaped by the ways in which energy is put to work in service of commerce, growth and the everyday need to make a living. Worldwide and across town, uneven geographies of economic development and standards of living reflect underlying inequalities in the availability and cost of energy and how it is used. Whether in farming, manufacturing, retail, logistics or any other economic sector, the availability, reliability and cost of energy influences what, where and how things get done. The evolution of today's global economy, characterised by accelerating trade and the geographical separation of economic production and consumption, reflects important shifts in fuels and technologies of transportation. Trading and investment in energy and energy services (heat, power, light, mobility) are also economies in their own right: oil, for example, is the world's most valuable traded commodity and energy companies are among the largest in the global economy, while powerful technology firms – like Google and Apple – are increasingly important actors in the energy sector. Energy means work for millions of people worldwide, from fuelwood harvesting and gasoline retailing to coal mining and reactor safety specialists. In many parts of the world, energy-gathering activities occupy a substantial part of people's daily routines.

factor of production
a term used by economists
to describe key inputs used
in the production of goods
or services. Classical
economics identifies three
factors of production: land,
labour and capital. This list
can be expanded by
'unpacking' one or more of
the factors: the influential
political economist, Susan
Strange (1988), for example,
adds technology and energy
as fourth and fifth factors
(both of which can be
considered forms of capital),
highlighting how these have
substantially raised the
productivity of land and
labour.

Elsewhere, relatively few people are employed in delivering energy services, but the economic necessity of reliable and continuous energy supply creates a political opportunity for workers in the sector to leverage their position to improve pay and conditions. As a consequence, energy workers are frequently subject to 'disciplining' by business, government and the effects of technological change.

Overall, energy availability shapes the structure and form of economies at a range of scales, from the horizontal footprint and vertical skyline of cities, to the possibilities of workplace democracy and the globalisation of manufacturing. This chapter considers two different dimensions of the relationship between energy and economy. It looks first at energy's role as a classic **factor of production** and input to economic development. It explores the changing nature of the correlation between energy consumption and economic growth, before considering the challenges facing economies that are highly dependent on energy exports. Second, it considers energy systems as an economic 'frontline' in a century-old struggle over whether economies should be organised around private or public forms of ownership.

ENERGY, GROWTH AND UNEVEN DEVELOPMENT

The ability to call on progressively larger quantities of energy – and to access energy in the high-quality forms of liquid fuels and electricity – has long been a measure of development. The relationship between economic growth and energy is complex. There are broadly two different lines of enquiry, depending on whether energy consumption or energy production is made the independent variable. The former is associated with debates around the energy intensity of economic growth; and the latter with the challenges of resource-based development, which are particularly significant for countries with a high degree of dependence on energy exports (see **Chapter 4**).

The energy intensity of economic development

energy intensity
a measure of the energy
efficiency of a country's
economy, based on units
of energy input per unit of
GDP output. A country with
a higher energy intensity uses
more energy per unit of GDP
output than a country with
a lower energy intensity.

The relationship between economic output and energy consumption can be described as the **energy intensity** of economic activity (see also **Chapter 6**). Influences on energy intensity include the type of energy resources consumed, technologies of energy conversion, prevailing prices for energy, and the structure of the economy (such as the relative significance of heavy manufacturing vs services). Nonetheless, five general patterns can be identified. First, economic output is positively correlated with energy input at the national scale so that, as a general rule, countries with a higher level of GDP consume more energy than those with lower levels of economic development (Figure 2.1). This 'basic rule' around energy intensity – that countries with a high level

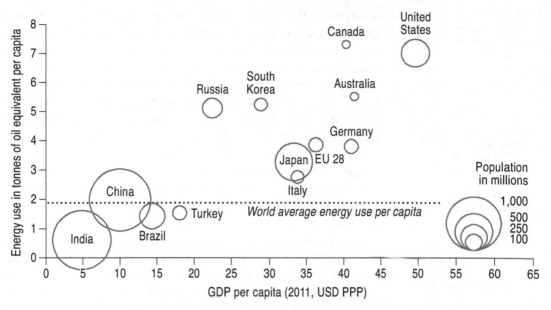

Figure 2.1 GDP per capita versus energy consumption per capita (based on European Environment Agency 2016)

of economic output are more energy intensive than those at a low level of economic output – has been recognised for a long time, although its limitations are also clear. Figure 2.1 indicates countries with a similar level of GDP often consume widely different quantities of energy: some countries are far more 'efficient' than others in converting energy use into economic output, and there are also geographical factors like size and location at work. Furthermore, beyond a certain point additional energy consumption no longer delivers further gains in GDP. In addition, the national averages used to construct a graph like Figure 2.1 conceal very wide internal variations in energy consumption, and the co-existence (often in the same town or city) of high and low energy economies.

A second pattern is that the energy intensity of many economies on a per unit of GDP basis has declined over time. Over the past 200 years energy intensity has fallen in Western economies by at least a factor of three, and possibly much more (Ruhl et al. 2012; BP 2017). However, there is some debate around the long-term trend: whether or not it holds true depends on what one counts on the input side (Gales et al. 2007). Analyses that focus only on commercially traded fuels suggest an inverted-U pattern: energy intensity rises from a low base during the early phases of economic development (reflecting the growing role of markets in the allocation of energy sources as well as a shift in economic structure from agriculture to industry) to a peak and then subsequently declines (due to improvement in energy efficiency and, in some contexts, further evolution in economic structure towards services). Ruhl et al. (2012), for example, indicate energy intensity peaking in the UK in the 1890s and declining by a factor of four over the following century, with the US, China and India peaking in the 1920s, 1980s and 1990s respectively. It is important

OECD/non-OECD

The Organisation for Economic Co-operation and Development (OECD) is an organisation of thirty-five member states. Founded in 1961, the OECD has evolved into a broad organisation that seeks to promote the growth and governance of market economies and democracy. It includes some of the largest energy consuming economies in Europe, North America, South America and Australasia, although China, Russia, India, Brazil, South Africa and Indonesia are not members. The distinction OECD/non-OECD is frequently used in energy statistics: the two groups of countries have different energy profiles (see **Chapter 4**), and it provides a short-hand for identifying the global shift

to remember that energy intensity is a measure of energy consumed per unit output. An economy improves its energy intensity if economic output grows faster than energy consumption, but the absolute amount of energy consumed continues to grow. There is also a tendency to focus on commercial forms of energy, but studies that also include traditional fuels, fodder and food in their analyses challenge the assumption that energy intensity was low in the pre-industrial period (Gales et al. 2007). These studies indicate not a 'peaking' in energy intensity but a longer-run decline, driven primarily by technological change and the substitution of higher-quality energy sources.

A third pattern is that energy intensity has not only declined but, across a number of countries, it has begun to converge so that as national economies grow they become more alike in their energy intensity. Recent data indicate energy intensity worldwide declined by nearly a third between 1990 and 2015, a pattern not limited to the global North (e.g. **OECD**) but also seen in most **non-OECD** countries (EIA 2016). Fourth, as economies evolve and become more dependent on the service sector and information technologies and data centres, as well as the 'cool chain' in food supply, so the quality and continuity of electricity supplies

Figure 2.2 Energy's multiple economic landscapes (clockwise from top left): World trade, Singapore container port (photo credit: cegoh); Suburban sprawl, Nevada, USA (photo credit: USDA Natural Resources Conservation Service); Heavy industry, aluminium smelter in Kazakhstan (photo credit: Yakov Fedorov); Electrical power transmission and distribution (photo credit: Google Data Center, The Dalles, Oregon by Visitor7/Wikimedia)

become more important as the economic consequences of a 'black out' are far more significant (see Figure 2.2). Fifth, the growing availability of high-quality energy sources and a general decline in the energy intensity of economic activity have not occurred everywhere equally. In fact, some economies have reduced domestic energy-intensive forms of activity and now import those energy intensive products from elsewhere. Consequently, the significance of energy as a factor of production is geographically uneven. This geographical diversity has important implications: it underpins patterns of economic integration (spatial differences in energy availability and energy intensity influence international energy trade, the distribution of certain forms of manufacturing, and territorial carbon emissions, for example); and it explains why the 'global energy dilemma' takes distinctive forms in different parts of the world (Bradshaw 2014).

in energy demand away from Western, industrial economies towards the so-called 'rising power' economies.

ENERGY AND ECONOMIC GLOBALISATION: OVERCOMING THE 'FRICTION OF DISTANCE'

Today's global economic landscape is characterised by an unprecedented degree of spatial interaction through international trade and investment, and widespread international travel. The scale of these movements reflects the availability of cheap fossil fuels and a set of key technologies – notably marine diesel engines and gas turbines in aviation – by which energy is converted into useful work. The lower energy density of low-carbon, renewable energy sources (see **Chapter 1**), and the relatively strong linkage between global economic interaction and the enhanced spatial mobility of goods and people via fossil fuels present major challenges for a low-carbon energy transition (see **deglobalisation**).

Energy is an important factor of production, along with land, labour and capital. A long-run decline in the cost of energy relative to these other factors has enabled a widening and deepening of economic globalisation (Sorrell 2009). The relative fall in energy costs over time is linked to improvements in the productivity of production processes, which have made it possible to produce more output for a given energy input. A growing supply of higher-quality energy inputs, and falling prices for energy services, enabled manufacturers to expand energy use over time, substituting capital equipment (electric motors, internal combustion engines, steam turbines) for the more diffuse energies provided by water, wind and human and animal muscle. The progressive substitution of high-quality (i.e. concentrated) forms of energy in production has underpinned the remarkable increases in labour productivity that characterise twentieth-century economic growth (Figure 2.3). The replacement of humans by robots on manufacturing production lines is a visible example of this trend.

deglobalisation
a term used to describe the prospect of a 'reversal' of the post-war trend towards global integration via international trade. The re-shoring of some manufacturing activity to the United States is often cited as an example, encouraged by falling energy and raw material costs (due, in part, to shale gas development, see **Chapter 4**). More generally, deglobalisation describes a situation in which global trade grows more slowly than global output: this has been the case in recent years, and is a reversal of the post-war experience of increasing economic integration as international trade grew much faster than total output.

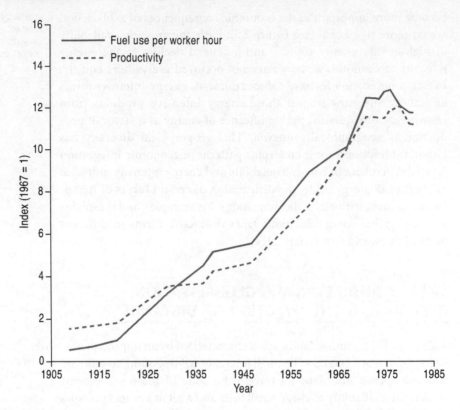

Figure 2.3 Labour productivity and energy use over time (based on Hall et al. 1986)

The overall effect of the transition towards higher-quality energy sources – what Smil (2010) terms the 'grand fuel sequence' (see **Chapter 9**) – has been a general loosening, over time, of the need for industry to locate close to sources of energy supply. A consequence is that other factors – notably the cost and productivity of labour – have become important as determinants of industrial location. The 'global shift' in manufacturing over the last forty years – associated with the de-industrialisation of traditional manufacturing heartlands in Europe and North America, and the rise of new 'workshops of the world' within Asia (Dicken 2011) – may be traced, in part, to this decline in the relative significance of energy as a factor of production and a much greater sensitivity to the cost of labour. It also explains why non-OECD countries now consume more energy than those in the OECD, and why over 90% of energy demand growth between now and 2050 will come from non-OECD countries.

Trade is a second way in which energy systems have contributed to more globally integrated economic landscapes. A defining characteristic of economic globalisation is the way international trade has grown faster than economic output: while the production of goods increased six-fold between 1960 and 2010, international trade in these goods increased twenty-fold (Dicken 2011).

At the world scale, the value of trade now accounts for over 51% of GDP, compared with 24.5% in 1960 (there is, however, wide variation at the level of individual countries). Behind such figures is the mundane business of moving goods around: the economic history of globalisation – of accelerating economic integration via trade – is, in part, a history of the application of energy in water, rail, road, and air transportation which has enabled a process of **time–space compression**. The contemporary global economy, characterised by extensive international trade, has taken form over the last half-century around a particular suite of transportation technologies (diesel engines and gas turbines, in particular) whose function depends on inexpensive and plentiful supplies of liquid fossil fuels. These engines are "fundamentally . . . more important to the global economy than are any particular corporate modalities or international trade agreements" (Smil 2010: 18).

Energy historians trace a series of innovations in fuels and energy technologies (**prime movers**) that, over time, have been associated with distinctive waves of economic integration: water power, sailing vessels and road building associated with early mechanisation in the 1770s; steam power, railways and shipping in the 1830s; electricity and steel in the 1890s; and automobiles, highways, aircraft and airports in the 1950s. Smil (2010), for example, comments on developments in construction, design and rigging of sailing ships from the seventeenth century onwards that enabled more effective conversion of the wind's kinetic energy to forward motion, enabling an increase in the speed of movement and expanding the scope of long-distance ocean trade. It was developments in steam propulsion, however, that enabled the development of inter-continental bulk cargo shipping (e.g. cotton, copper and wool) and commodity market integration in the first half of the nineteenth century. The greater speeds afforded by steam power, and the capacity to transport large tonnages at lower cost, drove an uneven process of spatial integration. In the United States, for example, expanding rail networks served to integrate regional commodity markets for wheat, meat and forest products, so that wide regional variations in price for these goods began to disappear: for example, the difference between wheat prices in New York City and Iowa fell from 69% to 19% between 1870 and 1910; similarly, the spread of wheat prices between Liverpool and Chicago fell from 57.6% in 1870 to 15.6% in 1913 (O'Rourke and Williamson 1999). The application of steam power to transportation increased both the intensity and extent of economic integration. It drove further the geographical separation of production and consumption to the continental scale, expanding significantly the 'supply zones' feeding energy and raw materials to manufacturing economies in parts of Europe and North America.

New transportation technologies and fuels accelerated 'time-space compression' in the early twentieth century, consolidating the decisive nineteenth-century shift in economic integration. The transition from coal to oil as the

time–space compression
a term widely used by geographers to describe the processes that cause relative distances between certain places to fall, as measured by travel time or cost. The availability and cost of energy are a significant influence on this process, although they are not the only factor. Time–space compression is not a uniform process: as some places become 'closer together' (through improvements in transport or the subsidising of transport costs for example), other places become relatively further apart, resulting in a dynamic and uneven economic landscape.

prime movers
a term used by energy historians and engineers to refer to the machines that, by converting primary energy into mechanical power, allow people, goods and materials to be moved around. Sails, steam engines, electric motors, diesel engines and gas turbines are important examples.

primary transportation fuel (first in shipping, then in road and rail) reflected the superior energy services that oil could provide, as a result of its liquid character and higher energy density. Oil's higher energy density changed the economies of scale required for crossing space, allowing the size of vehicle units to fall – from the train and tram to the automobile – and an increase in the power output for a given size or weight of engine. Oil's energy density enabled the evolution of the internal combustion engine (where oxidation/combustion on a small scale released a sufficient amount of energy to enable the direct movement of a piston), as opposed to the much larger, external combustion engines associated with steam power (Smil 1999). In cars and planes and on buses and trains, the number of passenger-kilometres travelled increased ten-fold between 1950 and 2000, rising from an average of 1,400 km per person to 5,500 km per person.

Where an earlier wave of globalisation expanded the geographical reach of economic interaction, the second half of the twentieth century has witnessed a deepening of globalisation as the intensity of international economic interactions has increased. Scale economies in transportation have accelerated economic integration by decreasing 'frictional costs': the speed of freight movements via land and sea has not changed substantially on a kilometre per hour basis, but significant time–space compression is apparent over the last fifty years when measured in terms of tons per kilometre per hour. Achieving economies of scale in marine transportation has rested on improved ship design, the introduction of containerisation and the scaling up of tankers and bulk carriers, increases in the size and efficiency of marine engines, and the continuing availability of inexpensive bunker fuels (prices for which are related to crude oil and currently carry no 'carbon tax'). Marine diesel engines, in particular, have facilitated a 'true planetary economy' (Smil 2010) via the inexpensive transportation of bulk raw materials and finished goods. The price of bulk shipping declined steadily in the second half of the 20th century, so that by 2004 it was half as much per ton as in 1960 (Hummels 2007: 142). The net result is that, for many goods, shipping costs are now a relatively small proportion of final product costs, so even relatively low-value goods can now be transported significant distance. The growth of global production chains – in which a series of discontinuous sites spread across the globe are functionally integrated in the production of a single product – reflects this reduction in the significance of transportation costs. In a similar way, falling air shipping costs have extended the transportable range for perishable, high-value agricultural products – such as fresh fruits, vegetables and flowers – enabling retailers (in the global North) and producers (in the global South) to take advantage of 'counter-seasonality'. The effect has been to collapse both space and time as it relates to global food provisioning: distant agricultural landscapes in the global South become tightly linked to the purchasing desires of consumers in the global North, while consumers' experience of seasonality is transformed via the 'endless summer' of fresh fruits and vegetables available through supermarkets (Freidberg 2004; Cook 2004).

The globalisation of manufacturing supply chains makes possible a geographic displacement of energy use and emissions, enabling consuming countries to effectively 'outsource' the demand for energy and emissions associated with all stages of product manufacture other than final consumption (Peters et al. 2011). The scale of energy and emission transfers via trade can be significant (Liu et al. 2010; Machado 2001). Research on UK trade, for example, shows how consumer demand has been increasingly met by importing goods into the country. Although there has been a fall in the UK's territorial emissions of greenhouse gases (i.e. emissions occurring within the UK), emissions associated with producing goods outside the UK and importing them to service UK demand have risen (Baiocchi and Minx 2010; Roelich et al. 2015). These import-related emissions are directly related to consumer demand within the UK, but a territorial system of emissions accounting (like that used by the United Nations Framework Convention on Climate Change) means they are attributed to countries outside the UK in which the goods are produced. Overall, the large net transfers of energy to industrial economies associated with economic integration through trade have ensured an expanding availability of cheap energy in these economies, enabling the gains in industrial productivity and improved standards of living described above. This relationship can be described as an 'ecological debt,' incurred by industrial economies and owed to the global South (Martinez-Alier 2002). It highlights the social and spatial transfers of energy and raw materials associated with deepening globalisation in the post-war period.

ENERGY SYSTEMS AS ECONOMIES: STATES, FIRMS, MARKETS, CONSUMERS

Power stations, pipelines, plantations and pylons are not just energy technologies or infrastructures. They are also economic assets. Owners of these assets manage and deploy them to create value. By asking who owns an asset, how the asset creates value for the owner, and who decides how the asset is used, it becomes possible to understand the economies created in and around energy system. This means focusing on flows of capital and the creation and capture of value, alongside the capture and transform of materials and the movement of electrons or fuels. Energy systems are owned and managed in a diversity of social forms, including public utilities, community trusts, state corporations and investor-owned corporations. In most countries, a variety of these forms will co-exist at any point in time, creating a complex organisational ecology that confounds simple descriptions of 'market' or 'state' organised forms. The United States, for example, is widely identified with the promotion of liberal markets and private capital: yet many electricity and gas assets are owned by city governments, the federal government is a major operator of hydro-electricity facilities, and the state shapes patterns of energy trade and investment in significant ways – for example, for forty years (up until December 2015) the federal government banned crude oil exports. Conversely, many nationalised energy companies will be part-owned by investors and listed on stock exchanges.

Several different 'trajectories' of change can be identified in regards to the ownership of assets. These reflect broader political and technological trends, and the continuum of state and market logics associated with different forms of capitalism around the world. While some patterns of ownership have endured for long periods of time, the last 100 years has also seen a pendulum-like motion between private and public forms of ownership across a continuum in the energy sector. Each of these trajectories has ideological overtones, as they reflect prevailing ideas about how and for whom economies should be organised. Yet none occurs in a vacuum, and they are best understood as responses to the opportunities of limitations of existing forms of social organisation, and shaped by the emergence of new technologies. We briefly describe four trajectories below: the development of public ownership models, privatisation and the promotion of liberal markets, a renewed interest in municipal forms of ownership, and the disruptive potential of the 'platform economy' associated with the hybridisation of energy and information. These models apply to the 'formal' economy, although in some contexts people more often turn to a parallel 'informal' energy economy to access energy (see above).

Public ownership

Public ownership is widespread in the energy sector and takes a variety of forms. Cuba's Ministry of Electricity, Norway's Statoil, the Los Angeles Department of Water and Power (DWP), and local energy co-operatives are examples of public ownership in the energy sector, although they imply different degrees of government control over financial and strategic decisions and responsiveness to end-user needs. In the first example, control over energy assets lies directly within the administrative structures of central government. Widely used in the former Soviet bloc, and in India and China until reforms of the 1990s, Ministries of Power were also a feature of many market economies until the 1980s: in New Zealand, for example, electricity was generated and transmitted by the New Zealand Electricity Department (NZED), which formed part of the Ministry of Energy. Electricity sector reforms in the late 1980s, however, converted NZED into a different form of public ownership: the government-owned corporation. Corporatisation separates operational assets from policy and regulatory issues, and introduces commercial structures and incentives. State-owned corporations can be wholly owned by the government (i.e. 100% of shares are held by the state), or majority-owned where the government retains at least a 51% stake and other shares are sold off and traded on a stock exchange. Around two-thirds of the shares in Statoil, for example, are held by the Norwegian state via the Minister of Petroleum and Energy. A third model of public ownership is the publicly owned municipal utility (**public utility**), in which power

public utility
a privately or publicly owned organisation that manages infrastructure for delivering an essential public service. A range of services are provided this way, including water, sanitation, gas, electricity supply and telecommunications: some broader definitions include airlines, railways and broadband communications. Where public utilities are privately owned, they are extensively regulated by the state: the New York Public Service Commission, for example, regulates electric, gas, water, and tele-communication industries in the state of New York.

generation and distribution services are owned and operated by local government (town, city or county) on a non-profit basis and on behalf of residents. There are over 2,000 publicly owned electric utilities in the United States serving around 48 million people, of which Los Angeles DWP is the largest. A fourth model is the energy cooperative in which assets are owned by its end-users (rather than by national or local government). A co-operative is a form of 'club' or private-partnership run for its members on a non-profit basis. It shares characteristics of both private and public ownership and is not reducible to either of them. Today, around 900 electricity co-operatives serve 42 million people in the US.

Historically there have been several 'waves' of public ownership in energy, including the municipalisation of power generation and gas and electricity distribution in Europe and North America in the inter-war years; national-isation of coal, gas and oil production, power generation and energy-intensive industries in Europe in the post-war period; the expropriation and national-isation of fossil fuel reserves and infrastructure previously owned by trans-national energy companies in the 1960, together with the assertion of 'permanent sovereignty' over natural resources (including oil, coal, gas and uranium) in the context of decolonisation and the bid by newly independent countries to establish a new international economic order; and a renewed interest in models of public ownership in the wake of a severe financial crisis (2008), growing concerns about energy security and the need to address climate change.

Key drivers of public ownership have been its potential for co-ordination and long-term planning (avoiding wasteful duplication and other forms of market failure); and the capacity of the sector as a strategic tool for implemen-ting economic and social policy goals, from rapid industrialisation and rural electrification to delivering environmental improvement and addressing regional inequality. Electricity and gas are classic examples of **natural monopolies** as both are network industries characterised by a high level of necessary expenditure on setting up and maintaining infrastructure (see **Chapter 3**). These fixed costs create very limited incentives for competition and, at the same time, lead to significant economies of scale. While natural monopolies can be either privately or publicly owned, many countries embraced public ownership of electricity, gas and other energy networks during the twentieth century. A large number of countries in Europe brought energy assets into public ownership and centralised control via nationalisation in the period following the Second World War. The UK government, for example, embarked on a series of nationalisations in the 1940s that extended to coal, gas and electricity (amongst other) sectors as well as many energy-intensive industries such as transport (railways, airlines), and iron and steel. This was not a universal experience, however: countries such as Japan, Germany and the United States retained private ownership in the electricity and gas sectors in the post-war period.

natural monopoly
an economist's term for a sector characterised by high costs of distribution, associated with building and maintaining extensive infrastructure networks. High fixed costs promote economies of scale in these sectors: distributing an additional unit of gas or electricity costs very little, but each additional unit sold means infrastructure costs can be spread more broadly. Sectors with these characteristics are therefore said to be 'naturally' monopolistic, because the service they provide can be delivered at the lowest cost by a single provider.

Energy's capacity for modernisation and development has led policymakers to view it as a one of the 'commanding heights' of the economy and critical tool for steering development. Many newly independent/decolonising countries sought a leading role for the state in the energy and power sector, incorporating these sectors into the administrative structure of government (as a Ministry of Power, for example), setting up national oil and gas companies and declaring national sovereignty over natural resources. The phrase 'commanding heights' is widely used in regard to public ownership of strategic economic sectors and first attributed to Lenin who, in the wake of the Russian Revolution in 1917, made electrification integral to the communist project of banishing poverty and building a classless egalitarian society. The National Five Year Plan, a key instrument of the centrally planned economy and synonymous with state socialism around the world, was first pioneered by the State Commission for Electrification of Russia (GOLERO) in 1920, which proposed the construction of thirty new regional power plants and electricity-intensive industries across the country (Figure 2.4). At around the same time, the United States also saw public ownership of energy assets as a key tool in driving economic development and addressing the failures of market capitalism (notably in the context of the Anti-Trust legislation in the 1920s and the Wall Street Crash of 1929). The federal government invested in a massive multi-purpose dam-building programme from the late 1920s until the mid-1960s under the Bureau of Reclamation. It continues to own and operate many of these assets, such as the Grand Coulee Dam on the Columbia River, and the Hoover and Glen Canyon Dams on the Colorado, so that today over 50% of the country's hydropower capacity is federally owned. The Tennessee Valley Authority, established in 1933 and combining flood control, electricity generation and industrial development, is the largest public power utility in the country and one of its largest producers of electricity. From the 1950s, the federal government's role in new dam construction became less significant, giving way to local municipalities and investor-owned utilities.

These waves of public ownership have created enduring legacies. Many of the most powerful energy companies are publicly-owned. The world's largest oil and gas companies – including Saudi Aramco, National Iranian Oil Company, PdVSA, Petronas, China National Petroleum Corporation, Rosneft and Gazprom, for example – are nationalised industries, either majority or wholly owned by national governments rather than private investors. Locally owned public power utilities, owned by residents and customers rather than share-holders, are an important part of the economic landscape in the United States and increasingly within Europe (see below). Publicly-owned electricity generation in the US, which also includes electricity from facilities owned by the federal government, accounts for nearly a quarter of the country's total electricity supply. Most of the electricity consumed in China is generated by five state-owned firms and transmitted across wires owned by two state-owned grid companies. The largest generator of electricity in Africa is South Africa's state-owned Eskom, which supplies over 95% of the country. In Canada, Australia and New Zealand, government-owned corporations have key roles in electricity generation and supply: BC Hydro, for example, is a Crown Corporation that owns hydroelectricity and gas-fired power plants and distributes power in British Columbia; and New Zealand's Meridian Energy, supplying around a third of

Figure 2.4 Propaganda poster outlining Lenin's vision for Soviet Russia (1924): 'Communism is Soviet power plus electrification of the whole country'

the country's electricity, is 51% owned by the national government. The UK's fleet of nuclear power stations, providing around 15% of the country's electrical power, are also state-owned, although not by the British government: they are owned and managed by EDF Energy, a wholly owned subsidiary of Électricité de France, which is majority-owned by the French government. Most countries now operate a 'mixed economy' in the energy sector, in which some elements of public ownership sit side by side with private (investor-owned) ownership.

Energy sector liberalisation and privatisation

deregulation
the removal of state controls on the operation of an economic sector. Energy sector deregulation in relation to electricity and gas, for example, centred on the removal of controls over price, the range of companies able to sell to customers, and the use of particular fuels (e.g. the prohibition of gas for large-scale electricity generation in the UK until 1991). Deregulation is part of a broader process of treating energy as a general commodity rather than a public good, and introducing market logics (such as competition and consumer choice) to guide energy investment, trade and consumption.

The energy sector has been a critical site of political-economic change in the last thirty years (Bridge and Bradshaw 2015). The introduction of market principles, commercial logics and private capital into the energy sector has been a key trend, although the direction of change has not been uniform. Dominated for much of the twentieth century by vertically integrated natural monopolies (both publicly and privately owned), the utility sector has been an experimental frontier for **deregulation** and the introduction of liberal markets and competition. At the same time, a legacy of public ownership has made the energy sector a target of privatisation initiatives. The electricity sector, in particular, has been a site for experiments in deregulation, which have sought to remove barriers to economic competition and expand the role for markets in determining where, by whom, and at what price electricity is generated and consumed (see **Example 1**).

Economic liberalisation policies have created a degree of convergence in market structures for energy, but geographic variation persists because of the way liberalisation policies articulate with existing national political economies. In some cases (such as Argentina), market reforms were rolled back while, in the UK, state intervention is an increasingly ubiquitous feature in the electricity market in the name of securing public goods such as energy security or decarbonisation: examples include 'strike prices' significantly above market rates for nuclear power and renewables (providing the generator with government-backed guaranteed prices) to encourage investment in low-carbon electricity generation, and 'capacity payments' to the owners of gas-fired power stations to ensure there is sufficient generating capacity to meet peak loads. Although Europe has pursued energy market liberalisation since the 1990s it has not been a universally shared experience, and state ownership of upstream and/or downstream parts of the production chain remains common. In several countries that embraced liberalisation – involving a rolling back of the state's ownership and control of energy assets and associated with the growth of cross-border energy trade and investment –concerns are now being expressed about import dependency and external control (Helm 2005). One of the legacies of energy market liberalisation, however, is that governments that embraced liberalisation now lack the capacity for a coordinated response to public concerns around energy security, climate change and the affordability of energy.

EXAMPLE 1: LIBERALISATION IN THE ELECTRICITY SECTOR

Liberalisation aims to relax direct government control and introduce market principles for allocating goods, capital and labour within an economy. Electricity was an early target of neoliberal economic reforms: the military government of General Pinochet in Chile introduced wholesale markets for electricity in 1978 and privatised the sector in 1982. Many other countries, provinces and states followed suit over the next two decades, particularly in the global South where energy sector privatisation and marketisation were a central plank of the 'Washington Consensus' on economic development policy. Comprehensive electricity sector privatisation was undertaken in the UK during the 1980s and early 1990s. In England and Wales, for example, the incumbent, vertically integrated state monopoly – the Central Electricity Generating Board – was 'unbundled': generation was separated from transmission and distribution, and competition encouraged in energy supply (Joskow 2008).

The liberalisation of gas and electricity markets has been a central EU policy objective for nearly two decades: liberalisation directives were first issued for electricity in 1996 and gas in 1998, and a more ambitious 'Third Package' of legislative proposals was adopted in 2007. In the United States, steps towards liberalisation were taken with the Public Utilities Regulatory Policy Act (1978) which, on the grounds of encouraging renewable energy, allowed new entrants into the gas and electricity sector. Liberalisation was significantly enhanced, however, by the Energy Policy Act (1992), which required owners of regional transmission lines to allow access to other generators of electricity (Troesken 2006). California deregulated its electricity market during the 1990s with the political objective of lowering consumer prices by increasing competition. Deregulation required vertically integrated, investor-owned utility companies, such as Pacific Gas and Electric, to divest their generating assets, and around half of California's generating capacity was sold to new market entrants. Utilities then purchased electricity from these generators in the wholesale electricity market, and distributed it for sale to customers in the retail market. Deregulation contributed to a severe power crisis in California in 2001, which was experienced as rolling blackouts and surging wholesale power prices. It culminated in the State of California bailing out bankrupt utilities like Pacific Gas and Electric, by purchasing power in the wholesale market on their behalf (Sweeney 2002).

Liberalisation and privatisation have had several consequences on the economic landscape. First, it has spurred a growth in transnational investment so that 'national' energy systems have become more spatially integrated and interdependent. Policies to liberalise energy trade and investment have driven a re-scaling of energy systems with energy companies, infrastructures and markets expanding beyond national borders (see **Chapter 4**). Between 1990 and 2011 the worldwide stock of foreign direct investment in the upstream (extractive) part of the energy sector increased eight-fold to US$ 1.4 trillion. Downstream investments in the energy sector (in electricity and gas services) increased by a much larger amount: the worldwide stock of foreign direct investment in electricity, gas and water services rose over fifty-fold to US$ 516 billion (UNCTAD 2013). This growing international interdependence of critical energy systems – experienced as changes to their territorial form, ownership and control – poses a major challenge for national energy policy. A key feature of the contemporary global energy economy is the emergence of new transnational corporate actors from outside Europe and North America, associated with a shift in the centre of gravity of the world economy towards Asia and the emergence of a multipolar world system (de Graaff 2011; Bridge and Le Billon 2017). While not limited to fossil fuels, this process is exemplified by the growing transnational activities of state-owned oil and gas companies such as PetroChina (China), ONGC-Videsh (India), Statoil (Norway), Rosneft (Russia), Petronas (Malaysia) and Petrobras (Brazil). Transnationalisation of these state-owned firms has gone hand in hand with their increasing integration with private companies (de Graaff 2011). The deepening globalisation of international energy markets, then, relates not only to the growing reach and intensity of trade and investment, but also to significant changes in its organisational structures. Overall, the net effect of this process of globalisation is that once-national systems of provision are increasingly porous, with "component parts of the world economy . . . increasingly interconnected in qualitatively different ways from the past" (Dicken 2011: 52). The implication is that energy systems are no longer 'nationally' contained but increasingly are shaped by decisions and interactions at multiple scales.

Second, liberalisation has enabled new entrants to come into the market. Liberalisation not only breaks up established monopolies but, more fundamentally, it also transforms what counts as relevant expertise for managing the sector and opens up new models of ownership and investment. Traditional utilities remain key players in liberalised energy markets but they are joined by banks, capital funds, and technology companies like Google, Apple and Amazon. Macquarie, for example, is a global investment fund headquartered in Australia, and a major provider of debt and equity capital to energy firms worldwide, as well as holding significant investments in energy infrastructure and energy trading: it is, for example, the largest non-producing trader of natural gas in the US. The US-headquartered tech-giant Google has invested in renewable energy projects around the world through a range of financial mechanisms and power-purchase commitments in pursuit of powering its data centres and offices with 100% renewable energy by 2025. Similarly, Apple has made large energy investments in China, including taking a 30% stake in China's largest wind-turbine manufacturer (CNN 2016). However, some of these consequences of liberalisation have been paradoxical: the liberalisation of the electricity sector in Germany, for example, led to a spate of mergers and acquisitions among previously separate (and mostly private) companies resulting in the emergence of 'the big four' (E.On, RWE, EnBW and the Swedish company Vattenfall) (Wagner and Berlo 2015).

CASE STUDY: ENERGY LANDSCAPES OF DIGITAL DATA – ARCTIC DATA CENTRES IN NORTHERN SWEDEN

**Contributed by Isaak Vié and Magdalena Kuchler,
Uppsala University, Sweden**

Sitting astride the Arctic Circle at the northern tip of Sweden lies the region of Norrbotten. Formerly known for its heavy extractive industries, Norrbotten is fast becoming host to a growing number of digital data centres. The arrival of the first European Facebook data centre in 2011 attracted the attention of other companies, drawn by the region's abundant energy supply, electrical infrastructure and cool climate. Firms with expertise in data centre construction and cloud computing technology have subsequently partnered with four municipalities in the region (Boden, Luleå, Piteå and Älvsbyn) to create the 'Node Pole', a regional development cluster to attract new clients. As of 2017, nine data centres have been established in Norrbotten hosting services for a wide range of international companies (BMW, Hydro66). Leading data-centre operators and specialist firms (Milestone, Fusion-io, EMC) have also set up facilities. Node Pole has recently been acquired by Vattenfall and Skellefteå Kraft, two prominent Swedish power companies.

A key component of the digital economy, data centres require large quantities of energy for processing and cooling. Data centres worldwide currently use around 3% of the global electricity supply, although their rapid development means the amount of energy they consume is doubling every four years (*The Independent* 2016). Norrbotten has an oversupply of low-carbon energy, sourced from hydropower plants combined with new developments of wind power (Vié 2017). Much of this infrastructure is a legacy of the region's energy-intensive industrial base in forestry, iron and copper mining, and steel production. Norrbotten is sparsely inhabited but produces an eighth of Sweden's electrical power, most of which is exported to southern parts of the country or regional neighbours. Data centres in Norrbotten have been able to take advantage of the pre-existing electrical network. The regional electricity grid is robust and stable, with contingency measures in place in case of outages. Tax regulations enable data centre operators to secure large discounts on electricity prices:as low as $0.0006 per kWh, prices are among the cheapest in the world (Vattenfall AB 2017). Innovative cooling methods have been developed to reduce energy consumption, which utilise the harsh climate of Norrbotten to maximum effect while also maintaining the required air quality parameters necessary to run electrical equipment. The combination of innovative practices has enabled new records to be set for efficient power utilisation.

Forecasts indicate an extra 200 data centres will be needed in Europe over the next decade or so, and Norrbotten alone could host up to 50 of these via the Node Pole cluster. The Node Pole model works because of a set of climatic and electricity supply conditions that are not necessarily found in other parts of the world. The region's abundance of cheap low-carbon electricity is also closely tied to broader energy choices in Sweden. The country's electricity supply relies heavily on hydro (41%) and nuclear (43%) sources, and it remains to be seen whether Norrbotten can continue to offer highly discounted electricity prices should Sweden finally pull the plug on nuclear power production.

A third consequence is the fundamentally uneven character of economic globalisation. The 'deepening' of globalisation may lead to some forms of geographical convergence because, for example, regional differences in energy intensity, technology choice, and regulatory standards become less pronounced. However, it also reproduces significant patterns of spatial difference (or produces them in alternative forms). The call for 'universal energy access' explicitly recognises that patterns of energy trade and investment have been remarkably 'sticky' over time: flows of energy – and flows of investment into energy infrastructure – remain highly concentrated and geographically uneven, reproducing large differences in per capita energy consumption at the global scale (see **Chapter 6**). Recognition of the wide differences in energy consumption between the global North and South drove inclusion of access to modern services within the UN's Sustainable Development Goals (the previous Millennium Development Goals had made no reference to energy). Geographical differences *within* countries can also be significant, particularly in parts of the global South where access to energy services often reflects a broader urban/rural divide. While patterns of international trade in modern commercial fuels like oil and gas have changed with geographical shifts in the centre of gravity of the world economy, the overall level of geographical concentration has not altered significantly. The international trade in coal, oil and gas remains dominated by a handful of major exporters and importers (the five most significant account for 60% of all oil imports; and 48% of exports). The geographies of connection and disconnection created via trade are politically and economically significant, and highlight the importance of thinking about 'national' energy policy as transcending the domestic sphere and fundamentally linked to foreign policy and international trade and development. From this perspective energy diplomacy is a significant part of national energy policy (Goldthau 2010).

Re-municipalisation

The dominant policy view on the organisation of energy sector in the past couple of decades has been that shifting assets to the private sector and creating markets in energy improves efficiency and responsiveness to consumers. The actual record of privatisation and economic liberalisation is more mixed, however. Since the financial crisis of 2008 (to which governments responded with massive state intervention, including nationalisation of parts of the banking sector) there has been a renewed interest in models of public ownership (Florio 2013). This has taken a wide range of forms (Haney and Pollitt 2013), but a significant trend has been a renewed interest in the municipal ownership of infrastructures for energy generation and distribution. Municipal ownership of gas and electricity services not only reintroduces public ownership but also re-scales it, from the national state to the administrative scale of the city (Cumbers 2012). Indeed, it is one of several alternative scales for thinking about public forms of ownership in energy which "tend(s) to oscillate between the non-state co-operative, the municipality and the nation state as the principle scales of governance" (Angel 2016: 2). Municipal provision has a long history and its return involves bringing back into public ownership assets that had previously been sold off and privatised, and/or replacing outsourced services with direct provision by local government. The drivers for remunicipalisation are diverse. They include growing acknowledgement of the costs and risks associated with private sector provision, and an appreciation of the 'public goods' character of energy services around issues like affordability, environmental sustainability and security of supply.

EXAMPLE 2: A 'DEATH SPIRAL' FOR LARGE REGULATED UTILITIES?

The growing role of renewables in the energy mix of many countries is causing a shake out of established business models in the energy sector. Europe's big utility companies are on the frontline of this process, particularly those owning nuclear, coal and gas assets. The market value of the top 20 energy utilities in Europe has halved since 2008, and Germany's E.ON and RWE have been among the most significantly affected. Squeezed in the power market by the growing penetration of renewables and falling wholesale prices for power, European utilities reduced the official value of their electricity generating assets by over 100 billion euros between 2010 and 2015.

The intermittency of many renewables presents a fundamental challenge to the model of a regulated utility supplying base-load power. In the case of solar PV, for example, supply peaks during the day enable solar operators to grab the juiciest part (i.e. highest prices) in the electricity market. In many parts of Europe, new renewable capacity has compounded a more general problem of over-supply and static demand. Utility companies over-invested in coal and gas generating capacity on the basis of projections of electricity demand growth that have not materialised (due to a combination of the falling energy intensity of GDP in Europe and a general economic slowdown).

The situation has been described by insiders as an 'existential crisis'. The Chief Financial Officer for RWE observed that "Conventional power generation, quite frankly, as a business unit, is fighting for its economic survival" (*Economist* 15 Oct 2013). Some see in the struggles of utility companies a new order coming into being, in which decentralised generation and municipal public utilities have a greater role. Others point to the scale of the investment challenge ahead, if society is to respond effectively to the four imperatives for energy system change (see **Introduction**), and the negative impact on investment of weakened utility companies.

The trajectory of remuncipalisation has been most pronounced in Germany where, between 2007 and mid-2012, more than 60 new local public utilities were established and over 190 electricity distribution networks returned to the public sector (Hall et al. 2013). There are also initiatives to strengthen the role of municipalities in energy trading markets and increase role in electricity generation. Hall et al. (2013), for example, describe the example of Trianel, a joint procurement initiative by over eighty *stadtwerke* (German municipal energy providers). Trianel was "formed in 1999 to facilitate energy trading by German municipal companies in the liberalised electricity and gas markets" (Hall et al. 2013: 204) but has grown over time to build power stations and wind farms (including the first municipal off-shore wind farm in the North Sea) and develop links with other municipally owned energy companies elsewhere in Europe. Germany's recent experience with municipalisation needs to be understood in the context of the country's exit from nuclear power, and its shift to renewables for production of heat and electricity: the so-called 'Energiewende' (see **Chapter 10**) (Moss et al. 2015). In Germany, as elsewhere, cities have emerged as an effective scale for addressing energy and climate concerns but municipalisation is also a response to the reluctance of the 'big four' utility companies to invest in renewables (see **Example 2**). Municipalisation, then, is part of a wider search for forms of 'economic democracy' that state nationalisation found hard to foster and whose possibility privatisation swept away (Cumbers 2012).

CONCLUSION

In this chapter, we have explored how economic landscapes are shaped in significant ways by the availability and cost of energy, and the technologies and social practices that guide energy consumption. The intensity of energy consumption varies widely, both between and within national economies. At the broadest scale, there is clear positive association between economic growth and energy availability, confirming how the under-consumption of energy services is a major structural problem for some of the poorest countries due to limited energy access and/or poorly functioning infrastructure. As economies evolve over time, their demand for energy changes along with the ways in which energy is used: industrialisation, for example, drives major increases in energy demand associated with energy intensive processes (like metals manufacturing) but also the growth of a mass consumer market. As industrial economies mature, however, the energy intensity of economic output declines over time as the contribution of heavy manufacturing to GDP declines. Across industrial economies, a process of convergence in energy intensity can be seen, aided by the diffusion of energy technologies and norms of energy use (around heating, lighting and mobility, for example). Important variations remain, however, and issues of energy cost, availability and reliability (particularly around information technology and data-processing) continue to be a key determinant of a region's competitive position and role within the global economy.

The chapter has also considered how energy systems are economies in their own right. Electricity networks and oil and gas extraction and processing infrastructures, for example, generate economic value through buying and selling a product, require investment and deliver returns on capital, and employ significant numbers of people. The scale and significance of many of these energy economies has meant they are frequently the 'frontline' in the century-old struggle over whether economies should be organised around private or public forms of ownership. Decarbonisation objectives and the mitigation of climate change have, in many instances, introduced a new dimension to this longer-term struggle over by and for whom energy economies are organised. They raise the question of what forms of ownership and governance are appropriate to the scale of the contemporary energy challenge: i.e. how to have energy systems that ensure access to modern energy services and deliver security, reliability and environmental sustainability, while advancing social justice (see **Introduction**). The renewed interest in municipal forms of ownership around electricity, gas and urban heat networks reflects this search for more effective and sustainable solutions. It is representative of a wider process of creative destruction at work within the economic landscapes of energy, associated with disruptive technological change and evolving social demands about the environmental and social performance of energy systems.

QUESTIONS FOR DISCUSSION

- 90% of world trade moves by sea and the speed at which these goods move (km/h) has not changed substantially over the last century, despite major changes in propulsion technologies and fuels: if speed has stayed the same, what has changed?

- How might the idea of 'splintering urbanism' be applied to understand transitions in the ownership of energy infrastructure?

- To what extent can post-war suburbanisation be described as 'oil-fuelled'?

- How does energy availability influence the temporal rhythms of economies, across a range of scales (diurnal, seasonal, annual, generational)?

- Take two contrasting economic activities/landscapes (e.g. corn field, data storage/processing centre, steel-making): consider similarities and significant differences in how energy is used in these activities; to what extent can you explain the geographies of each activity by reference to how they use energy?

ACTIVITIES/POTENTIAL RESEARCH PROJECTS

- Using a publicly available data source, identify the major energy flows associated with a national (or state) economy. Construct a flow diagram to represent the scale of these energy flows and the sectoral composition of energy demand. Use a national (base) map to show major points of import/export and summarise the main vectors of energy flow. Write a short report to summarise your findings.

- Who owns the infrastructures through which your daily energy demands are met? Construct a diagram that identifies the principal economic actors involved in energy provisioning. Consider the implications of this ownership structure for your ability as a consumer to influence decisions about investments in energy efficiency and new sources of supply.

RECOMMENDED READINGS

■ Blanchet, T. 2015. Struggle over energy transition in Berlin: how do grassroots initiatives affect local energy policy-making? *Energy Policy* 78: 246–254.

Berlin's electricity distribution grid is one of the largest in Europe. In the 1990s, the city's local government sold the grid to private companies. This article analyses subsequent conflicts over efforts to re-municipalise the urban electricity grid i.e. to bring it back into local government ownership. By telling the story of Berlin, the author examines the influence grassroots actors can have in energy transition, and shows how they can influence both the overall vision and implementation of urban energy systems.

■ Overland, I. 2016. Energy: the missing link in globalisation. *Energy Research and Social Science* 14: 122–130.

This article highlights the central role of energy trade in economic globalisation. It develops an index to measure the extent of this 'energy globalisation' in the period 1992–2011, focusing on oil, coal and gas. The author shows how energy trade has grown in extent and intensity, but also notes how the development of unconventional oil and gas (including via fracking) runs counter to this trend.

■ Ruhl, C., P. Appleby, J. Fennema, A. Naumov and M. Schaffer. 2012. Economic development and the demand for energy: a historical perspective on the next 20 years. *Energy Policy* 50: 109–116.

Countries vary in their energy intensity, defined as the amount of energy consumed for a given output of GDP. Energy intensity depends, in part, on the structure of the economy (manufacturing, services, agriculture), fuel mix and type of technologies used. This article examines the evolution of energy intensity over two centuries, and draws an optimistic conclusion about the effects of economic globalisation over the next twenty years. The authors show how energy intensities have improved: i.e. economies have become more efficient at converting energy inputs into GDP. They argue that increased fuel specialisation, and a continuing convergence in the sectoral and technological makeup of national economies as a result of globalisation, will accelerate improvements in energy efficiency.

■ Smil, V. 2010. *Prime movers of globalisation: the history and impact of diesel engines and gas turbines.* Cambridge, MA: MIT Press.

An insightful account of the role of two of the world's most important prime movers in enabling a modern global economy, characterised by the long-distance movement of bulk goods and planetary-scale personal mobility. Diesel engines and gas turbines are among the most powerful prime movers ever developed: the former dominate road, rail and maritime transportation; the latter are central to air transportation, power generation and the movement of oil and gas through pipelines. The author examines the historical conditions under which diesel engines and gas turbines emerged, their significance when set against other prime movers, and the reasons they are likely to be with us for some time to come.

REFERENCES

Angel, J. 2016. Towards an energy politics in-against-and-beyond the state: Berlin's struggle for energy democracy. *Antipode* 49(3): 557–576.

Baiocchi, G. and J.C. Minx. 2010. Understanding changes in the UK's CO2 emissions: a global perspective. *Environmental Science & Technology* 44(4), 1177–1184.

Blanchet, T. 2015. Struggle over energy transition in Berlin: how do grassroots initiatives affect local energy policy-making? *Energy Policy* 78: 246–254.

BP. 2017. *BP Energy Outlook.* Available online at www.bp.com/content/dam/bp/pdf/energy-economics/energy-outlook-2017/bp-energy-outlook-2017.pdf

Bradshaw, M. 2014. *Global energy dilemmas.* Cambridge, MA: Polity Press.

Bridge, G. and P. Le Billon. 2017. *Oil.* Cambridge, MA: Polity Press.

Bridge, G. and M. Bradshaw, 2015. Deepening globalisation: economies, trade and energy systems. In *Global energy: issues, potentials and policy implications.* P. Ekins, M. Bradshaw and J. Watson (eds). Oxford: Oxford University Press, pp. 52–72.

Cook, I. 2004. Follow the thing: papaya. *Antipode* 36(4): 642–664.

CNN 2016. Why Apple is investing in wind turbines in China. Available online at http://money.cnn.com/2016/12/09/technology/apple-wind-turbine-china-investment/index.html

Cumbers, A. 2012. *Reclaiming public ownership: making space for economic democracy.* London: Zed Books.

De Graaff, N. 2011. A global energy network? The expansion and integration of non-triad national oil companies. *Global Networks* 11: 262–283.

Dicken, P. 2011. *Global shift: mapping the changing contours of the world economy*. London: Sage, 6th Edition.

The Economist 2013 How to lose half a trillion euros: Europe's electricity providers face an existential threat 15 October 2013. Available online at www.economist.com/news/briefing/21587782-europes-electricity-providers-face-existential-threat-how-lose-half-trillion-euros

Energy Information Administration. 2016. Global energy intensity continues to decline, 12 July 2016. Washington DC. Available online at www.eia.gov/todayinenergy/detail.php?id=27032

European Environment Agency. 2016. Correlation of energy consumption and GDP per person. Available online at https://www.eea.europa.eu/data-and-maps/figures/correlation-of-per-capita-energy/

Florio, M. 2013. Rethinking on public enterprise: editorial introduction and some personal remarks on the research agenda. *International Review of Applied Economics* 27(2): 135–149.

Freidberg, S. 2004. *French beans and food scares: culture and commerce in an anxious age*. Oxford: Oxford University Press.

Gales, B. Kander, A. Malanima, P. and M. Rubio. 2007. North versus south: energy transition and energy intensity in Europe over 200 years. *European Review of Economic History* 11: 219–253.

Goldthau, A. 2010. 'Energy diplomacy' in trade and investment of oil and gas. In *Global energy governance. The new rules of the game*. A. Goldthau and J.-M. Witte (eds). Washington DC: Brookings Press, pp. 25–48.

Hall, D., E. Lobina and P. Terhorst. 2013. Re-municipalisation in the early twenty-first century: water in France and energy in Germany. *International Review of Applied Economics* 27 (2): 193–214.

Hall, CAS, C Cleveland and R. Kaufmann. 1986. *Energy and resource quality: the ecology of the economic process*. New York, Wiley and Sons.

Haney, A.B. and Pollitt, M.G., 2013. New models of public ownership in energy. *International Review of Applied Economics*, 27(2): 174–192.

Helm, D., 2005. The assessment: the new energy paradigm. *Oxford Review of Economic Policy*, 21(1): 1–18.

Hummels, D. 2007. Transportation costs and international trade in the second era of globalization. *Journal of Economic Perspectives* 21(3): 131–154.

The Independent (2106) Global warming: data centres to consume three times as much energy in next decade, experts warn. 23 January 2016. Available online at www.independent.co.uk/environment/global-warming-data-centres-to-consume-three-times-as-much-energy-in-next-decade-experts-warn-a6830086.html

Joskow, P.L., 2008. *Lessons learned from the electricity market liberalization*. Massachusetts Institute of Technology, Center for Energy and Environmental Policy Research.

Liu, H., Y. Xi, J.Guo, X. Li. 2010. Energy embodied in the international trade of China: an energy input–output analysis. *Energy Policy* 38(8): 3957–3964.

Machado, G., R. Schaeffer and E. Worrell. 2001. Energy and carbon embodied in the international trade of Brazil: an input–output approach. *Ecological Economics* 39(3): 409–424.

Martinez-Alier, J. 2002. Ecological debt and property rights on carbon sinks and reservoirs. *Capitalism Nature Socialism* 13(1): 115–119.

Moss, T., S. Becker and M. Naumann. 2015. Whose energy transition is it, anyway? Organisation and ownership of the Energiewende in villages, cities and regions. *Local Environment* 20(12): 547–1563.

O'Rourke, K. and J. Williamson. 1999. *Globalization and history: the evolution of a Nineteenth-century Atlantic Economy.* Cambridge, MA: MIT Press.

Peters, G., J. Minx, C. Weber and O. Edenhofer. 2011. Growth in emission transfers via international trade from 1990 to 2008. *Proceedings of the National Academy of Sciences* 108 (21): 8903–8908.

Roelich, K., J. Barrett and A. Owen. 2015. The implications of indirect emissions for climate and energy policy. In *Global Energy: issues, potentials and policy implications.* P. Ekins, M. Bradshaw and J. Watson (eds). Oxford: Oxford University Press, pp. 92–111.

Ruhl, C., P. Appleby, J. Fennema, A. Naumov and M. Schaffer. 2012. Economic development and the demand for energy: a historical perspective on the next 20 years. *Energy Policy* 50: 109–116.

Smil, V., 1999. *Energies.* Cambridge, MA: The MIT Press,.

Smil, V. 2010. *Prime movers of globalization: the history and impact of diesel engines and gas turbines.* Cambridge, MA: MIT Press.

Sorrell, S. 2009. Jevons' paradox revisited: the evidence for backfire from improved energy efficiency. *Energy Policy* 37: 1456–1469.

Strange, S. 1988. *States and markets: an introduction to international political economy.* London: Pinter Publishers.

Sweeney, J.L., 2002. The California energy crisis. *Journal of Energy Literature,* 8: 100–103.

Troesken, W., 2006. Regime change and corruption. A history of public utility regulation. In *Corruption and Reform: Lessons from America's Economic History.* Chicago, IL: University of Chicago Press, pp. 259–282.

Wagner, O. and Berlo, K. 2015. The wave of remunicipalisation of energy networks and supply in Germany: the establishment of 72 new municipal power utilities. In *First fuel now: ECEEE 2015 Summer Study;* 1–6 June 2015, Toulon/Hyères, France. Available online at http://nbn-resolving.de/urn:nbn:de:bsz:wup4-opus-59209

UNCTAD 2013. *World investment report 2013 – Global Value Chains: Investment and Trade for Development.* Annex Tables 24 and 26 (estimated world inward FDI stock by sector and industry, 1990 and 2011; estimated world inward FDI flow by sector and industry, 1990 and 2011).

Vattenfall AB. 2017. Energiskatter i Sverige, Available online at www.vattenfall.se/foretag/elavtal/energiskatter/

Vié, I., 2017. *Energy for information: the green promise of the Node Pole data centres,* Master Thesis, Department of Earth Sciences, Uppsala University, Sweden. Available at: http://uu.diva-portal.org/smash/record.jsf?pid=diva2%3A1110904

3 Infrastructural landscapes

- Appreciate the key characteristics of energy infrastructures and their variety of forms at different scales and in different local and national contexts.
- Understand the interrelated material and social dimensions of energy infrastructures and how these have developed over time and co-evolved with economic, political and technological change.
- Explain the emergence of high-density urban energy infrastructures and consumption patterns.
- Analyse how energy infrastructure breakdown can reveal the underlying spatialities and vulnerabilities of these infrastructures.

Connect. Provision. Circulate. Energy infrastructures link and divide peoples and places in ways that create distinctive landscapes of energy circulation. They bind energy consumers to energy suppliers and, at the same time, generate hierarchies of energy access and sharp inequalities between those connected and those who are not. The notion of 'infrastructure' is not precisely defined, and it can be hard to draw exact boundaries around where energy infrastructures begin and end. Conventionally, the category 'energy infrastructure' includes electricity transmission lines and substations; systems for distributing solid and liquid fuels, including oil and gas pipelines, storage tanks and pumping stations; and urban-scale networks for distributing heating and cooling services (normally in the form of heated/chilled water). In other words, energy infrastructures are connective (Shove et al. 2015) and circulatory (Edwards 2003), linking different places and, in simple terms, joining the supply of energy by generators and providers to the use of energy by consumers. Infrastructures enable and structure the distribution, movement and flow of energy so that it becomes available and accessible in particular forms (e.g. natural gas at certain pressure) and amounts (e.g. the power capacity of an electricity sub-station) at points of end use. The network character of an energy infrastructure is perhaps the key quality that the notion of infrastructure conveys, giving particular attention then to the geography of infrastructure and its distribution over space (Graham and Marvin 2001). There are other important qualities of energy infrastructures too. Star (1999) identifies a set of generic characteristics of infrastructure, including that it:

- *has permanence rather than needing to be reassembled each time it is used* – once established, infrastructures are, to some degree, sedimented in place, obdurate and 'locked in' physically (see **path dependency**), but also through their often deep embedding in the ongoing functioning of society;
- *develops through modular increments rather than being built all at once* – infrastructures such as gas pipeline networks or railway systems are typically extended over time, reaching across and into more spaces, but through that process prioritising some areas and supported functions over others;
- *tends to be hidden and forgotten but becomes visible upon its breakdown* – as long as infrastructures function effectively they are very much taken for granted and in the background, less so when they fail.

path dependency
a term that rationalises and explains why certain technologies, processes or practices continue to be used, even when they may be demonstrably inefficient or out-dated. The historical development of energy infrastructures has, over many decades, laid down established practices and rigid technologies that cannot be easily challenged or moved away from. For example, replacing the UK's centralised energy infrastructure with community-based systems of provision would be challenging both technologically and in terms of energy governance and control.

As we shall see, each of these characteristics are relevant to developing our understanding of energy infrastructures and their significance in wider energy systems in both conceptual and more practical terms.

Figure 3.1
Power lines and pylons as connective infrastructure, Iceland (photo credit: Gordon Walker)

We can push beyond the conventional definition, however, in thinking about the infrastructural landscapes critical to energy and society relations. On the supply side, a range of energy conversion technologies are also forms of infrastructure (Goldthau 2014). These include power stations (converting chemical to electrical energy) and their 'upstream' networks of fuel supply (e.g. the chain of uranium mining, transport and processing that precedes nuclear power generation; Karlsson 2009). However, these supply-side conversion technologies are the subject of other chapters in this book (see **Chapters 1 and 9**), and we do not discuss them further here. On the demand side, we can include a wide range of infrastructures that have important roles in shaping how energy is consumed, and energy demand built and sustained. For example, transport infrastructures – roads, rail networks, canals, airports – and their relation to the spatial forms of urban settlement have been particularly significant in the creation of energy dependencies and growing levels of global energy demand. The second half of the chapter explores this broader sense of infrastructural landscapes by focussing on the embedding of energy demand in urban settings.

In the discussion that follows, we explain how energy infrastructures are both material and social in character. We emphasise the constitutive links between the physical manifestations of energy infrastructures in (and below) the landscape, and the social processes through which these come to be, along with how they are owned, managed and operated. The chapter then explores the different scales at which we can analyse energy infrastructures, from those that are 'off grid' and localised, through to transnational networks of production, distribution and consumption. Using the scale of the city as a case study, the chapter then examines how energy infrastructures have become embedded in urban physical and social fabrics in the twenty-first century, in which specific systems of provision have developed to deliver energy, and introduce dependencies, in particular ways. The chapter ends by considering how these systems of provision and associated dependencies can lead to vulnerabilities in both the resilience of supply and of individuals and communities who rely on continuously available cheap energy. We highlight how energy infrastructures relate to social and economic contexts, and how these are crucial to the particular ways in which energy is delivered and consumed.

ENERGY INFRASTRUCTURE AS MATERIAL AND SOCIAL

Energy infrastructures evidently take a material physical form. They are constructed and built and have permanence, incorporating many different forms of intermediating material technologies (power lines, petrol pumps, underground gas storage reservoirs and so on). They, therefore, have a physicality and a scale that are important to their geography, and to how they are positioned in landscapes and stretch across space (see Figure 3.1). This physicality, in turn, shapes exactly how interconnections are made between different places and between suppliers, distributors and consumers.

Figure 3.2 Energy infrastructures become 'visible' when sudden failures occur in the system: a filling station without fuel in the UK as a result of an industrial strike by road tanker drivers (photo credit: Mike1024)

The immediate materiality of energy infrastructures (such as electricity pylons) can be important to the protest and contestation that sometimes erupts around development projects that are seen to conflict with or threaten valued place identities or senses of well-being (see **Chapter 7**). But energy infrastructures can also be largely invisible, hidden underground or submerged within other built structures. They can also be so much part of the accumulated historical evolution of everyday spaces and places that they are taken for granted and go unnoticed. As Edwards (2003: 185) comments about infrastructures in general 'they reside in a naturalized background. As ordinary and unremarkable to us as trees, daylight and dirt'. This does not mean that this less visible materiality cannot become controversial (the underground can still be a space of contestation, for example in relation to fracking and radioactive waste storage). As Star (1999) makes clear, the taken for granted can become suddenly very immediate and urgent in the event of failure – witness the drama and chaos of electricity blackouts or brownouts (see below), gas leaks, or petrol shortages (Figure 3.2).

This emphasises that energy infrastructures are enormously varied in their particular material makeup, scale, form and impact and we will explore this variety further in the next section. As a general point, these characteristics are to some degree reflective of the particular materiality of the energy flow that the infrastructural system is enabling. Electricity distribution, for example, needs and enables infrastructure that gas distribution does not. The materiality of electricity and gas has a key influence over the form of the infrastructure through which it is supplied. Electricity consists of electrons continually flowing in a

current

electrical current is the basis for supplying energy to appliances and other technologies that rely on electricity for their power. Electrical current is a charge carried by a flow of electrons in a wire and delivered via circuits. For the most part, electrical systems are based on alternating current (AC) where the direction of electrical charge varies; however, some systems (e.g. batteries or solar cells) use direct current (DC) where the charge is uni-directional. The distinction is less significant now, but in the late nineteenth century there was considerable debate over the safety and efficiency of AC and DC as rapid expansion of transmission networks forced decisions over which system to adopt.

peak load

for a given geographical area, electricity consumption can be defined in terms of the aggregate 'load' it imposes at a given point in time on the production and distribution system. Peak load refers to periods of time when, within this spatial context, aggregate consumption increases to a high level and the production system has to operate at full power in order to meet this load. This is usually anticipated and managed by grid operators on a daily basis, but concerns over peak demand overstretching capacity or relying on high carbon power sources have increasingly made it a political concern.

current and needs conducting wires to enable this movement; and to achieve a certain voltage and power needs certain capacities and designs of insulators, capacitors, transformers and so on, for it to be safely and efficiently distributed.

The materiality of electricity is also distinctive: electricity cannot be stored on a large scale (Osti 2016), so that flows through the system need to be effectively instantaneous, providing a real-time connection between electricity supply and its use. In large-scale electricity networks (or grids) there is a *spatial* dislocation between production and consumption (it happens in different places often very distant from each other), but *not* a dislocation in time. This means that storage technologies are not currently part of the normal makeup of the electricity distribution network (although new storage methods are emerging), and that the electricity system is particularly sensitive to failure in any part of the connection between production and consumption, or inability to meet what is demanded (see later discussion of **peak load**). By contrast, gas (along with petrol, oil, coal and wood) can be easily stored. It is produced both at another place *and* at another time (Osti 2016): it sits waiting to be used in pipelines, and can be contained in bulk in large-scale gas tanks and storage reservoirs (in turn, raising safety and environmental concerns of a distinctive character). Material differences in energy forms mean, therefore, that their infrastructural systems are configured differently, with implications for how these infrastructures are made and maintained and for the challenges involved in their development and operation.

While the materiality of energy infrastructures is obvious, it is important to appreciate they do not *just* consist of technical 'hardware' – the material and physical structures or component parts of energy distribution networks, such as pipes, wires and petrol tankers. It is a basic tenet of contemporary socio-technical approaches to studying such technologies that these are always both material *and* social – or technical *and* social (Bijker and Law 1992). Social here has a broad meaning, and refers to the substance of human activity, ideas and knowledge systems, discourses, institutions and ways of designing, arranging and using, which are continually shaping and being shaped by technical artefacts. The key idea is that infrastructures embody interrelated social and technical dimensions that cannot be readily untangled and made distinct. This way of thinking foregrounds five significant analytical observations:

- Energy infrastructures are always designed, constructed and extended in their reach and scale through social processes that are situated in particular settings. Ideas, ideologies and expected uses and users are embedded or 'sedimented' within their design and development; they are 'cultural artefacts' (Hughes 1983: 2).

- Energy infrastructures are always owned, managed and operated, and are subject to rules, regulations and governance measures. They have a 'software' and are 'socially organised' (Walker and Cass 2007) in ways that reflect patterns of power, authority, expertise and political judgement.
- Infrastructures are integral to 'systems of provision' (Fine 2003) that connect production, distribution and consumption in a chain of activities that is specific, to some degree, to the form of commodity (or energy) and the cultural setting in which it is embedded.
- Infrastructures do not simply and innocently interconnect energy supply to pre-existing energy demands, but rather they can be actively involved in *creating* new energy demands and dependencies (van Vliet et al. 2005).
- Energy infrastructures are essential to economic change and development: they connect supplier and purchaser, enabling the selling of produced energy and the buying of useful energy to provide energy services. They are socially and materially intrinsic to capital accumulation and to the functioning of energy economies (see **Chapter 2**).

These are complex ideas and two examples – one historical and one more contemporary – can provide an initial opening up of their substance and implications. The first example relates to the historical development of electricity distribution infrastructures in the US, which has been the subject of detailed studies by geographers and historians (Hughes 1983; Howell 2011; Nye 1999). In a particularly fascinating example, Harrison (2013a, b) traced the development of electrification in North Carolina, focusing on the production of an 'uneven landscape of electricity' that has been a major contributor to energy poverty and uneven economic development in the region. Power plants and electricity distribution networks were laid out beginning in the early twentieth century by both public and private utilities, employing ideologies of 'progress and emancipation' to justify their investments and promote high levels of electricity use by consumers in order to generate hefty economic returns. This strategy of energy infrastructure development proved successful until the 1970s when various challenges to profits emerged and a foray into nuclear power left major power utility debts that had to be passed onto consumers. The outcome has been a combination of high local electricity prices and high levels of elec-tricity consumption that have proved enormously problematic for both businesses and low income consumers. Harrison comments that "the streams of electrons that pass through the distribution wires, and the prices that are charged for the use values they provide, are representations of past labour processes, past accumulation strategies and past power struggles" (ibid.: 473), emphasising the way that currently operating infrastructures always carry something of the politics and the social conditions at the time of their production. In another powerful example, Harrison (2013a) shows how the detail of the initial laying out of the electricity grid was strongly racialised, privileging white neighbourhoods and holding back electrification in pre-dominantly black areas. In short, the geographies of electricity infrastructure and consumption were starkly imbued with the racialised politics of the American South at the time, in ways that were simultaneously social and technical.

The second example is of developing patterns of natural gas use in emerging metropolises in the global South. Verdeil et al. (2015) examine the last twenty years of urban gas infrastructure development in three cities – Cairo (Egypt), Istanbul (Turkey) and Sfax (Tunisia) – all undergoing rapid processes of development and social change. They demonstrate how the emerging geography of natural gas distribution is partly about the proximity of natural gas reserves, but also reflects ambitious municipal economic development strategies, and programs aiming at cutting energy subsidies that have traditionally been used to support universal energy access for poorer consumers. The affordability of connection and access to this new infrastructure has become politically contentious. A growing middle class population is keen to modernise and expand energy consumption, increasingly leaving others behind. These tensions create major implications for governance, not only of the energy system "but, as protests related to energy services observed before and more strongly since the Arab Spring have shown, of national stability itself. Access to modern energy forms has become a major stake" (ibid.: 244).

Figure 3.3
Electric vehicle charging points: an example of new infrastructures creating new demands for electricity (photo credit: Gordon Walker)

Both of these examples have focused on how infrastructures come to be, expanding their geographical reach over time in ways that are subject to cultural and political ideas and economic conditions. They also demonstrate the unevenness that can permeate energy infrastructure provision. Energy becomes available and accessible for some but not for all, with patterns of provision reflecting wider currents (in these cases) of class division, racial discrimination, state restructuring and demands of capital accumulation. Both cases also demonstrate how infrastructure development can go hand in hand with the active creation of new uses and users of energy, a process that continues today, for example in the promotion and development of electric cars and new charging infrastructures (see Figure 3.3).

INFRASTRUCTURAL SCALES AND SPATIAL FORMS

The ways of thinking about energy infrastructure discussed so far open up much complexity. There is value, therefore, in identifying a scheme that can systemise some of the main characteristics of different infrastructural forms. The categorisation that follows therefore provides a way of distinguishing between infrastructural systems that particularly emphasise their geographical scale, extent and size – from the very immediate, to the local, national and transnational. As summarised in Table 3.1, each of these material scales can also be characterised in terms of their related social and institutional characteristics – how they are owned and managed, the ideas and meanings they carry and the governance arrangements that can be applied. While this is a rather conventional hierarchical approach we shall see that these infrastructural scales and forms:

Table 3.1 Different scales of infrastructural system and their characteristics

	Spatial extent	Ownership and management	Ideas and meanings	Governance
Self-provision and off-grid	Very local and immediate	Private	Right to energy, self-reliance, frugality, sustainability	Minimal, may be incentivisation schemes for development
Local network	Community, neighbourhood or city scale	Municipal, community or private	Localness, sustainability, autonomy	Predominantly local or community
National grids and networks	Nationwide or regional	Public, private or public-private	Efficiency, control, universal provision, competition	National government and agencies
Transnational networks	Internationally expansive	Private, public, or public-private	Efficiency, competition, cooperation	Co-operative between nations

- can co-exist in space and time and work together in parallel and in tandem. They are not in this sense discrete alternatives;
- do not sit on a simple spectrum from undeveloped to developed, or basic to advanced. They rather fit in different ways into processes of development, change and transition;
- are general types, within which there will be much variety and potentially crossing between scales.

Self-provisioning and off grid

At this scale the provisioning of energy is very local, such that any involvement of infrastructure in movement and distribution is minimal if non-existent. Production and consumption have no need of external or extended interconnection, given that they happen in the same space and through the work of the same actors. A traditional example would be the self-collection of wood and biomass for burning in fires or stoves (see **Chapter 6**), which is still a significant part of energy use in poor, rural parts of the global South and found in other rural areas around the world on a small scale. Such arrangements often reflect a level of intense poverty in situations where infrastructures that can provide energy in other forms, or that enable trade and exchange rather than self-collection, are not available or accessible. In more wealthy and infrastructurally developed settings, self-provisioning can, however, also be driven by an active sense of self-reliance and resilience, an intent to be 'off-grid' and disconnected from commercial systems (Vannini and Taggart 2013). Purposeful 'frugality' (Evans 2011) and commitments to living in sustainable ways can also be significant, and associated with contemporary and future oriented processes of sustainability transition. Here, forms of technologically advanced local microgeneration (solar panels, pico and micro-scale wind and forms of biomass conversion) have become increasingly prevalent as a means of living 'off-grid' (in both global North and South settings), or more widely for having both a local immediate supply of energy, and a connection to larger scale grid infrastructures – a form of 'co-provisioning' (van Vliet et al. 2005).

feed-in-tariffs (FITs) a widely adopted mechanism to incentivise and accelerate the adoption of distributed renewable energy technologies. FITs typically include guaranteed access to the grid for new generators, a long-term contract for supplying electricity and a price for the electricity generated which reflects the higher costs involved compared to established coal, gas or nuclear power stations. FITs have successfully enabled diverse technologies (wind, solar, biogas, etc.) to be developed and secure investors a reasonable return.

Co-provisioning has rapidly grown with recent attempts to increase generation from renewable sources (Mendonça 2007). The emergence of both household and community forms of energy generation have occurred in part in response to policy instruments, such as **feed-in-tariffs**, that have created incentives for the installation of renewable technologies, such as photovoltaic cells and solar water heating (Walker and Devine-Wright 2008). But local action has also been driven by a range of environmental, economic and social motives, and by different governance arrangements. There are, for example, significant differences between the governance models adopted in the UK and Germany (see **Case study**).

CASE STUDY: PROMOTING COMMUNITY ENERGY IN THE UK AND GERMANY

Community energy projects involve collective action to reduce, generate or purchase energy. The UK and Germany have experienced substantial growth in electricity micro-generation as communities and individual householders elect to invest in renewable technologies, such as wind turbines and solar photovoltaic panels. Germany, however, has seen a much more rapid uptake of community-scale initiatives than the UK. Research by Colin Nolden (2013a; 2013b) in the southwest of England and in Germany explored how different scales of energy governance and regimes of innovation affect the deployment of renewable technologies. Using qualitative methods, he undertook interviews and observational research with renewable energy providers and community energy groups to examine the effectiveness of the different regimes in place in the UK and Germany.

The UK system of promoting small-scale renewable energy deployment has been governed centrally, and has relied on a nationally set, universal feed-in tariff that provides investors with a fixed payment per unit of energy exported to the electricity grid. It also depends on banking and lending arrangements that, in comparison to Germany, are relatively centralised. Uncertainty about the level and longevity of the feed-in-tariff in the UK has caused investors to lose confidence, so that deployment of small-scale renewables has been sporadic. By contrast, Nolden (2013b) found that Germany's success in promoting community energy projects has been achieved through what he terms a 'co-evolution' of technological development, energy policy, bank lending practices and institutional learning. Importantly, regional lending banks in Germany have provided investment for community scale energy initiatives.

Nolden's research highlights how very different outcomes can emerge in energy infrastructures through different scales of economic policy, planning and governance. His work shows the effects of path dependency on the development of energy infrastructures: in the UK, this has led to renewables deployment being adapted to and accommodated within a relatively centralised energy delivery system; while, in Germany, the multi-scalar and distributed nature of energy governance has created significant opportunities for community-based renewables investment. The scaling of political governance in Germany through the federal system has enabled different forms of regime to emerge, while the UK's centralised system of governance means that regions have little or no influence over issues like energy policy. Comparative research like this reveals how there are cultural and political challenges, as well as technical hurdles, in deploying small-scale renewables. The UK has a much weaker heritage of community-scale innovation and collective working than Germany. Supporting and promoting community energy in the UK requires, therefore, challenging cultural assumptions about the spatial scale and organisational forms in which energy is captured and distributed, including those embedded in prevailing policy frameworks.

Local networks and supply systems

district heating networks
a system deploying the economies of scale that can be gained from having one, central heat source and distributing heated water or steam to a number of properties (domestic and non-domestic) and/or streets of buildings. They are reliant on the use of highly insulated pipe work to transport hot water and / or steam and can be fuelled by a wide range of both low-carbon and carbon-intensive fuels. In many parts of Northern and Eastern Europe district heating systems are well established, whereas in other contexts district heating is seen as innovative in part because of changes in the relationship between consumers and providers, removing the need to have a heat energy source (e.g. gas-fired boiler) in each property.

distributed energy system
an energy system that reflects the changing interactions between infrastructures, technologies, scales and practices in the twenty-first century. Conventional energy systems are often centralised, linear and inflexible and only responsive to a relatively minor degree. Distributed systems of energy provision and distribution use a diverse mix of energy sources, storage mechanisms, and distribution systems and demand control mechanisms to manage efficient energy delivery, often using 'smart' technologies like mobile devices to enable consumer control.

Local infrastructural networks whose spatial extent is limited to a particular community, neighbourhood, complex, urban area or region, have historically been the first stage of energy infrastructure development. Examples include local trading networks of solid or liquid fuels, through to 'town gas' networks supplying a limited urban area and local electricity grids similarly spatially bounded and contained. In the early twentieth century in the UK electricity distribution took the form of a mass of separate local electricity grids with their own generating stations and operating company. In 1930 in the south-west England and Wales region alone there were 165 separate utility companies supplying electricity to their customer areas (Hughes 1983: 360). As noted in the North Carolina case study discussed earlier, these network developments were associated historically with new technological innovations and urbanisation processes. But ideas of municipal provision of new energy services to citizens (light, cleaner forms of heat generation), sometimes aligned with profit-making by new supply and distribution companies, were also important. In the global South local networks continue to be a feature of early-stage infrastructural developments, including local community scale electricity networks set up as part of rural electrification projects powered by solar panels.

Local networks are not, however, just a first stage of development to something larger and more sophisticated. In settings where the underlying geography is fragmented – such as a country like Indonesia (Chelminsky 2015) made up of hundreds of separate islands or in remote areas such as the Brazilian Amazon (Gómez and Silveira 2012) – locally bounded energy networks may well be seen as the optimal arrangement. Local **district heating networks** that distribute heat to local housing, public and commercial buildings are long established in Scandinavia and on a major scale in Iceland (see **Example 1**), and are being actively developed in the UK and other countries as part of policies driven variously by sustainability, fuel poverty and energy security concerns (Webb 2014). There are also examples of local electricity micro-grids being established, including by private sector organisations such as Apple, who are designing their major Campus 2 development in Cupertino in California (a 176-acre site for 13,000 employees) to be able to operate independently of the wider electricity network (Hering 2014). In these ways, local energy networks are part of moves towards **distributed energy systems** that are far less centralised and monolithic, with implications for how they are run, managed and organised. Again, however, these co-exist with other scales and forms of energy infrastructure.

EXAMPLE 1: DISTRICT HEATING SYSTEMS

District heating is well established in parts of Scandinavia. In Iceland it works on a massive scale, with over 90% of buildings in the capital city, Reykjavik, heated using a system for distributing geothermally heated water (Lund et al. 2008). Because of the volcanic activity of Iceland, there is much potential for tapping underground heat, both in the form of naturally hot water in springs, pools and aquifers, and through installed systems for heating up cold water using high temperature steam. From the 1930s onwards, pipeline distribution systems have been developed in different parts of the island, providing space heating and a variety of other uses including swimming pools and snow melting. The energy use profile of the country has consequently been transformed. The transition in heating to geothermal sources and away from burning coal, oil and turf means the country now has a heating system that is both self-sufficient and low-carbon. There are over 200 district heating networks in Iceland, from the very large to the very small, with the length of the distribution network extending to over 4,000 km.

National grids and networks

The notion of a national grid is historically particular and limited to certain countries around the world and to particular forms of energy provision – electricity and gas, most specifically. However, national infrastructure networks have also, in effect, existed (at least for periods of time) for coal, oil and other fuels where these have been self-contained without significant trade and imports across national borders. Various studies have tracked in some detail the way that ideas of developing a national electricity grid emerged from earlier patterns of locally or regionally distinct networks, driven by ideas of technological efficiency and advancement, along with notions of 'universal provision', public ownership and control and (sometimes) with wider 'nation-building' ambitions (Hughes 1983).

In the UK there was a step change in infrastructural scale and form when the Electricity (Supply) Act of 1926 was passed that created the 'national gridiron', latterly to become known as the National Grid. As Horrocks and Lean (2011) explain, such a move was designed to effectively manage electricity production and distribution, albeit still reliant on a scattering of independent energy providers. In due course, the national grid became the responsibility of the Central Electricity Generation Board (CEGB) with progressive centralisation eventually leading to the nationalisation of both electricity production and all distribution in 1947/48. This form of unified publicly owned network governance was taken apart, however, with the privatisation of the electricity industry in 1989 (see **Chapter 2**). This involved separating out electricity generation companies from the National Grid who own and manage the main grid network, the regional distribution network operators (DNOs) who run the local networks to customers, and the electricity suppliers from whom customers buy their electricity. This complex and shifting history demonstrates that there is not one form of ownership or governance arrangement that is inherent to a national-scale grid infrastructure.

Transnational networks

Where energy flows cross borders we need to recognise the transnational nature of infrastructural networks. We might most immediately think in these terms of relatively recently developed fixed infrastructures, such as transnational gas pipelines or electricity interconnectors. However, there are earlier histories of trade in fuels – patterns of export and import of wood, coal, oil and gas – by ship and road transport, that have long constituted extended networks of infrastructure for the movement and distribution of energy across space. As discussed in **Chapters 2 and 4**, international trade in energy has been enormously significant economically and politically, and the infrastructures that enable this – marine oil and liquefied gas terminals, port coal handling facilities through to undersea and overland pipelines – are an integral part of that trade. The particular phenomena of transnational grids – that provided interconnections across national boundaries and between separately run and managed national or regional grids – has been particularly important in Europe. As well as the many interconnecting electricity and gas lines that run across the European mainland, newly imagined and proposed transnational networks continue to proliferate, such as the 'Northern Atlantic Energy Network' (Orkustofnun et al. 2016) connecting between Arctic, Nordic and Northern Regions to link renewable energy capacity to major energy markets.

Energy infrastructures are evidently enormously varied. While they are in some ways stable and obdurate, they can also clearly evolve considerably over time, following varied directions of change and co-existing in different forms. Gas for cooking and heating can, for example, simultaneously be distributed to end users by pipeline networks, by trucks and in individual canisters, forming a set of infrastructural networks that connect together, but to some degree operate independently of each other. Examining these different characteristics of energy infrastructures and the scales at which they are configured emphasises their successful functioning as part of the distribution and delivery of energy to end users. However, as argued earlier, infrastructures do not just end with supply and distribution, but are very much also implicated in the creation and sustaining of patterns of energy demand. Accordingly, the following two sections explore the relationship between supply and demand as embedded within particular systems of provision, and the specific vulnerabilities that emerge when these systems fail to function and provide as intended.

ENERGY INFRASTRUCTURES AND ENERGY CONSUMPTION IN URBAN CONTEXTS

Urban contexts provide the most striking and significant spaces within which densely packed infrastructures and high levels of energy demand come together (Rutherford and Coutard 2014), with consequences for both the carbon footprint of major cities and the stresses on energy systems that derive from rapid processes of urbanisation and city development. In this section we focus on such urban settings, extending our analysis through to the 'fingertips' (van Vliet et al. 2005) of infrastructural systems and the technologies they power.

If you live in a country with hot summers and at a reasonable level of income, have you considered how much your familiar everyday world is dependent upon air conditioning? As Shove et al. (2014) highlight, the adoption of mechanical cooling systems has become increasingly prevalent around the world, embedded in buildings and vehicles but also in the practices that make up the flow of everyday life – driving, watching TV, eating out, sleeping in hotels to name but a few. Implications have followed, for example, for the timing, location and performance of work, leisure and domestic activities, through to the clothes that are worn in climate-controlled settings and patterns of eating and drinking. Explaining the emergence of air conditioning, both historically and in the contemporary world, is not entirely straightforward. As Shove et al. (2014: 1517) note, it is necessary to follow the "differently evolving contours of the various, but sometimes overlapping practices into which mechanical cooling is integrated". This reveals to some degree processes that are globalising air conditioning in a very standardised way through, for example, internationally reproduced approaches to modern office or car design, but also processes that reflect the particularities of local cultures of living with heat. What has been fundamental, however, has been the development of electricity infrastructures that have the capacity to power air conditioning installations en masse and at scale; to the point that in some urban areas nearly all internal space, of whatever form, is artificially cooled. In other words, the systems of provision of cooled indoor air have the production, distribution and use of energy at their core.

Although for many in high income countries the availability of electricity is now accepted and expected as a basic utility, it is only in the relatively recent past that energy infrastructure has developed in such a way that a reliable, on-demand system of provision has become available to the majority. As already explained, in such countries electricity and gas provision has typically evolved from local settlement-scale infrastructures, through to larger-scale networks with a centralised energy production and distribution system. This process of extension, consolidation and centralisation has had profound consequences for relationships between people and energy infrastructure in the urban context (Graham and McFarlane 2014).

In North America and Europe after World War II, the development of integrated and centralised energy infrastructures was associated with the

Figure 3.4 Gas cookers have transformed domestic food preparation: an advert from *The Ladies Home Journal* (1948)

introduction of new household technologies that reduced the burden of domestic chores (Figure 3.4). The relationship between supply infrastructures and consumption norms evolved differentially and at different speeds from setting to setting. In the UK many urban households did not have access to natural gas supplies until the 1980s, and so home heating was often provided by a single coal fire or solid fuel boiler. Such technologies were not a huge distance away from traditional forms of home heating and cooking, which were reliant on labour-intensive technologies like kitchen stoves, individual fires or log burners. Making provision for heat, cooking and hot water was time intensive and factored into daily routines, so that some energy-dependent activities were related to particular days of the week (e.g. doing the laundry on Monday) reflecting their labour-intensive character and the strongly gendered social organisation of the household.

Jump forward to the present day and the majority of households in the UK (and other wealthy nations) now have access to technically advanced heating, hot water and cooking fuel sources that require little or no labour to provision or control to a high degree of precision. Modern central heating systems enable users to programme temperature settings to within 0.5°Celsius over multiple time periods per day and over weeks of a year. Likewise, gas and electric cooking appliances provide instant heat, while automatic washing machines and refrigerators provide energy services that once took up both time and space in household budgets. Changes like these, then, have enabled and co-evolved with shifts in the ways we live our lives. Shove et al. (2012) have explored these issues in depth, examining the dynamics of everyday social practices – shared routines that develop over time and across space (see **Chapter 6** for more discussion) – making clear how these dynamics are intricately related to the affordances offered by centralised energy provisioning (Shove 2003). As energy has increasingly become readily available, it has enabled changes in shared practices and expectations of normal ways of living. These changes, in turn, have contributed to increases in energy demand and shifts in the patterns of how energy is used.

Another striking example of how the ubiquitous availability of energy has become structured into urban landscapes is provided by the flows of cheap oil and petrol that have fundamentally shifted patterns of transport and mobility across the urban world (Figure 3.5). Technological developments in motor transport that enable motor vehicles to efficiently travel long distances have been aligned with major investments in attendant infrastructures that support this technology: petrol stations, motorways, service stations, bridges, tunnels and elevated roadways. These major investments in infrastructure have

Figure 3.5 The iconic and sprawling infrastructure of the American Interstate freeway system (photo credit: Michigan Department of Transportation)

wrought huge impacts on the built environment, and changed everyday practices of work, leisure and dwelling, resulting in what Urry (2004) termed a culture of 'auto-mobility'. For example, over 70% of Americans now live in 'auto sub-urbs', large sub-divisions of land put down to extensive suburban developments, where driving in an absolute necessity for living (i.e. shopping, going to school, work, leisure). Once conceived as edge-of-city havens away from the squalor of urban poverty during the industrial revolution (Baritz 1989), the provision of cheap oil has afforded North Americans a means by which to have the benefits of urban civilisation while living in a highly stylised version of a rural idyll (Baldassare 1992). Indeed, between the 1930s and 1960s, the invocation of suburban culture as a highly desirable middle class way of life witnessed the development of particular and prescribed forms of suburban culture, which were deemed to be signifiers of success (Baldassare 1986; Baritz 1989; Clapson 2003; Miller 1995). Moreover, suburbia afforded, through the motor car, the achievement of key political goals for a free-market society. As Rajan (2006: 113 notes):

> Automobility, on its part, has become the (literally) concrete articulation of liberal society's promise to its citizens that they can freely exercise certain everyday choices: where they want to live and toil, when they wish to travel and how far they want to go ... Its constitutive visual image is one of dignified convoys of individual cars, vehicles whose solitary drivers can remain separated from each other as they collectively pursue private goals on public highways. As such, this picture captures the salient features of cars in a post-Enlightenment order: the experience of driving, identified by the quiet pleasures of the open road, speed, power and personal control, neatly complements the functionality of covering distance, managing time and maintaining certain forms of individuation.

lock-in
describes how a dominant technological design, fuel type or energy system maintains its position and resists potential alternatives. The sources of lock-in can be both technological and institutional, with the two often reinforcing one another. Classic examples in relation to energy are the late nineteenth-century struggle between AC and DC current, the internal combustion engine, and nuclear reactor design (see **Chapter 9**). The concept of lock-in can also be applied to carbon-intensive, fossil fuels

Such a pervasive cultural norm as suburban living and the freedom afforded by car driving is therefore key to understanding how energy and carbon intensive mobility has become such a valued and essential part of everyday life for so many people. Kunstler's (1994) *The Geography of Nowhere* popularly characterises the ways in which auto-mobility '**locks-in**' individuals and sections of society into ways of living that can only be sustained by personal motor transport, which can itself only be viable with constant supplies of cheap oil.

The energy-dependent patterns of living we have examined show how energy infrastructure enables and promotes new forms of energy consumption, and how energy infrastructures can become deeply embedded in daily routines and practices in ways that are then hard to move away from. Every upward shift in demand for an energy service (like cooling, for example) further entrenches social expectations about its availability and perceptions of need. Enacting meaningful change in demand is, therefore, frustrated by deep **path dependencies**. For example, economic logics for shifting away from automobility and towards investing in public transport/transit systems do not fit well with already-established patterns of settlement characterised by low-density suburbs, where potential transit users are lightly spread across large areas. Enabling significant shifts towards cycling and walking is similarly spatially problematic.

(see Unruh 2012) where it describes the system-wide forces (e.g. subsidies for their production and consumption) that sustain their dominance and create barriers for alternatives.

Policy norms can also be caught up in path dependencies in ways that limit options for change. The energy policy analyst Catherine Mitchell (2008) argues that tackling environmental threats like climate change is unlikely to be achieved through what she terms the Regulatory State Paradigm (RSP), in which nation states utilise the free market to incentivise particular forms of energy provisioning, regulating when there are examples of market failure. She argues that such an approach can create 'ideological **lock-in**', where paradigms of policy making create and reinforce institutions that share the same assumptions and goals of the regime, thus stifling radical change. 'Options' for policy making are then narrowly defined, focusing on the use of the free market, the maintenance of existing and enhanced levels of 'service', and the perpetuation of an economic growth model founded on high energy inputs. As such, questions about 'needs' and how they have become established and alternative ways of provisioning quality of life are sidelined, in favour of a securitising approach that upholds reliability and affordability as key criteria for success. Mitchell's analysis provides the basis for unpacking some of the assumptions that have been made about systems of provision, and enables us to understand how such systems are intricately related to the wider political-economy of nation states and regimes of governing. It is also through this lens that we can begin to view and appreciate the ways in which highly centralised and demand-heavy systems of provision can come under strain and create vulnerabilities.

ENERGY FLOWS AND DISRUPTIONS: REVEALING VULNERABILITIES

The type of expectations of 24/7 continuity and ever-growing dependency on energy as an ingredient of a whole host of everyday practices, puts even more emphasis on concerns about disruptions in supply and the vulnerabilities this exposes. As discussed earlier, infrastructures can be seen as revealing themselves at moments of breakdown or failure (Edwards 2003), and the nature of centralised and highly complex systems of provision presents a variety of vulnerabilities that can have far reaching spatial and political consequences (Chaudry et al. 2011). We can categorise the vulnerabilities associated with infrastructures and associated systems of provision under the following headings: resource and capacity failures, technical disruptions, political instability and natural disasters.

demand response
the objective of reducing (or sometimes increasing) electricity demand for particular time periods in response to pressures on the supply system. These pressures may be capacity, cost and/or carbon related. Responsive reductions in demand typically occur during peak periods, while increases in demand are sought when there is an 'over-supply' e.g. on a windy and sunny day with a lot of low cost wind and solar generation. Shifting demand in time, making it flexible and responsive, can be achieved at a grid scale through time of use pricing and contracts with big energy users that pay them to cut their energy use in response to signals from the grid operator.

battery storage
conventional energy infrastructures are based on meeting demand as it occurs through the management of generation capacity, such as rapidly increasing fuel use in a gas-fired power station or releasing water in a hydroelectric plant to instantaneously generate electricity. Yet, concerns over energy shortages during peak periods of demand is leading to more investment in battery technology to store energy. This is being conceived at the domestic scale (to store energy from micro-generation plants) as well as 'battery farms' to provide large-scale input into energy grids.

Resource and capacity vulnerabilities have become key concerns in recent years, particularly for those focused on maintaining the resilience of energy systems to meet demand (Chaudry et al. 2011). Maintaining sufficient supplies of raw materials to generate electricity, keep gas flowing and ensure transport systems can work smoothly, continues to be a priority for nation states. Avoiding resource shortages and supply interruptions is, therefore, an essential concern and underlies much current policy making. This need to guard against resource supply disruption is matched also by concerns about peak load within the electricity system in particular, where peaks in demand from consumers can lead to capacity overload and threaten or cause blackouts across grid networks. Traditionally, energy providers have managed capacity through being able to switch between different forms of energy supply, but this relies on these being instantly accessible and able to meet both high demand in short-term contexts (e.g. in an evening) or for longer time periods (e.g. during a cold winter). Alternative ways of grid balancing through **demand response** are now being developed in which reductions in demand are incentivised during peaks and/or periods in which supply is constrained (Torriti 2015; Walker 2014). And most recently, developments in **battery storage** are being trialled to explore the potential of battery farms to deal with peak load, for example in the United States and Australia (*The Guardian* 2017a, b).

Second, is a concern with the vulnerabilities associated with technical failures of large and complex transmission systems. As discussed earlier, transnational networks have been established within the remit of institutions like the European Union that enable the transfer of energy between nation states and even across continents. However, inter-linked transmission networks present significant vulnerabilities because of their complexity. Van der Vleuten and Lagendijk (2010) undertook an analysis of the November 2006 European blackout, in which approximately 15 million households were left without electric power and twenty European nations were affected. The trigger for this event was a routine shutdown of one high voltage line in northern Germany to permit a cruise ship to pass along the Ems River. The resultant overloading of other lines led to a series of safety trips that rapidly (within seconds) had a domino effect across Europe. Their historical analysis of the antecedents to this technical failure highlighted the largely *ad hoc* nature of electricity grid integration across Europe from the 1950s, with regulation only emerging years later. Indeed, they argue that it is only recently that politicians have been willing to consider Europe's electricity grid as 'vulnerable', as opposed to the previously dominant discourse of 'reliability'. Accordingly, apparently technical vulnerabilities can also be viewed as ones that attain exposure through the shifting politics of energy.

Third, the geo-politics of energy vulnerability is an enduring issue that has had exposure since the middle part of the twentieth century (**Chapter 4**). In the oil crises of the 1970s, Arab oil exporting nations restricted supplies of crude oil to Western markets because of their support for Israel at a time of political conflict and turmoil in the Middle East. Subsequently Western nations, in particular the United States, have been strongly focused on securing reliable oil supplies and reducing the vulnerabilities associated with shifting energy geographies. Indeed, more recent conflicts have been explicitly associated with America's demand for oil, particularly in popular discourse (*The Guardian* 2014). The geography of energy conflict is focused on both those places where key resources are held (for example, the Middle East); and where there are dependencies between key economic and political blocks (for example, between the Russian Federation and the European Union), particularly erupting at times of heightened political tension (see **Chapter 8**) (Bradshaw 2009). What such conflicts demonstrate is the vulnerability associated with import dependency and limited diversity in energy sources, both of which can readily lead to supplies being cut off.

Finally, infrastructural vulnerabilities can emerge from naturally induced disasters, which can have both immediate impacts on transmission and longer-term political consequences that result in major changes in energy policy. At the most basic level, weather-related phenomena like major storms, flooding, extreme cold weather and heat waves can put pressure on energy systems and cause infrastructure to become disabled. For example, ice storms in eastern Canada in 2013 left thousands of households without power for days (CBC 2013). In the UK, winter storms and flooding in 2013 were responsible for thousands of homes being left without power during the Christmas holiday period, leading to a political backlash against power companies who were perceived to be 'slow to respond' (BBC 2014). **Example 2** describes a similar storm-related experience and the electricity dependencies it revealed.

These and other examples illustrate how power cuts caused by natural events can rapidly become commentaries on the vulnerability of energy networks, and lead to new regulations and expectations being placed on energy providers. A dramatic example is provided by the earthquake and subsequent tsunami in Japan of 2011, which casued floodwaters to disable the Fukushima nuclear power station and cause a partial meltdown. This led not only to a loss of supply, and debilitating and dangerous radioactive contamination, but also a reappraisal by many nation states of the vulnerabilities associated with nuclear power (see **Chapter 9**). The collective memory of the Chernobyl nuclear disaster of 1986 in the Soviet Union invoked by Fukushima resulted in several policy changes, including Germany's decision to abandon its nuclear energy programme. Questions were raised in many other nations over the safety and security of nuclear power, which has been a long-standing concern of opposition groups (Bickerstaff et al. 2008). In these ways, as Castán Broto et al. (2014) argue, infrastructures 'shocks' can catalyse systems change if the learning that develops involves critical evaluation of existing provisioning principles and their suitability for complex situations.

EXAMPLE 2: POWER BLACKOUTS AND ELECTRICITY DEPENDENCY

In a history of blackouts in America, energy historian David Nye (2010: 33) argues that blackouts are "carved out of the normal flow of time", creating "a new kind of social space". "The response to a power failure changes from one blackout to the next," he argues, "revealing how society is becoming more and more dependent on electric power". Such dependency and disruptions to normal life were revealed by an extended power failure in the city of Lancaster (UK) in December 2015, caused by the major flooding of an electricity substation (see DEMAND Centre (2016) and Royal Academy of Engineering (2016) for detailed accounts, reflections and analyses). While the 'lights went out' and moving around and carrying on as normal 'in the dark' therefore became problematic, much else stopped working that was less immediately obvious and visible. Payment systems in shops and restaurants shut down, mobile communication networks stopped functioning, and students at Lancaster University could not stay in their accommodation because fire alarms were not working. Many of these direct and knock-on consequences were experienced, often without easy or effective adaptations being available. Ordinary life, its flow, rhythm, pathways and expectations of convenience were substantially disrupted, in ways that were substantially more extensive and significant than people's memories of power cuts in previous decades.

CONCLUSION

Energy infrastructures take many different material forms. While they are primarily thought about in terms of energy distribution – the 'connective tissues' of grids, pipelines, petrol distribution, heat networks and similar – the notion of infrastructure can also be to taken to include technologies of generation and end consumption. In this chapter we have shown that however they are defined, all forms of infrastructure are both social and material in their constitution and functioning, and can be configured at very different scales – from the local off-grid to the expansively transnational. Energy infrastructures are established and governed, developed, added to and sometimes transformed in ways that are necessarily politically determined, with outcomes that will have uneven consequences and implications for different places, communities and economies – including those arising from infrastructural failures and breakdowns. In the context of climate change, energy's infrastructural landscapes both lock-in high energy ways of living and offer possibilities of change towards a low carbon future.

QUESTIONS FOR DISCUSSION

• What are the key generic characteristics of infrastructures, and how can these be observed specifically in energy infrastructures?

- Do supply infrastructures evolve to meet energy demands, or do they also create new patterns of demand?

- What are the strengths and weaknesses of a more distributed and multi-scaled pattern of electricity generation?

- Evaluate the case for infrastructures that enable more extended and cross-border geographies of energy flow.

ACTIVITIES/POTENTIAL RESEARCH PROJECTS

- Undertake an audit of your daily activities by keeping a simple diary from when you get up in the morning until you go to bed. What energy sources and infrastructures are you dependent on? For example, if you have a glass of orange juice with your breakfast, what energy-providing and dependent infrastructures were required to produce, transport, sell and store the juice?

- Consider a scenario for the future where there is no electricity supplied to your home for 48 hours in the winter time. Which of your daily activities would be affected by this breakdown in energy infrastructure? How would this affect you and how would you cope in response to this external shock? Think about all the activities that might be affected: for example, it is likely an electricity blackout would mean a gas heating boiler could not function.

RECOMMENDED READING

■ Gómez, M.F. and S. Silveira. 2012. Delivering off-grid electricity systems in the Brazilian Amazon. *Energy for Sustainable Development* 16(2): 155–167.

In Brazil, millions of people have benefited from a rural electrification program involving the extension of the electricity grid. This paper argues that a new off-grid model is now required to further extend access to electricity to remote areas in the Amazon region. This includes giving attention to the need for more local and site specific solutions using renewable energy technologies that will enhance the inclusion of remote areas in universal access goals.

■ Mitchell, C. 2008. *The political economy of renewable energy.* London: Palgrave Macmillan.

This book explores the ways in which energy can be understood from a political economy perspective and highlights how various forms of political and economic lock in operate at various scales to promote particular kinds of energy pathways.

In the case of renewable energy, Mitchell demonstrates how existing scales of decision making, regulation and levels of public engagement lead to particular outcomes. Such path dependencies, as highlighted in this chapter's case study, mean that changing energy systems is highly complex and often very slow.

■ Rutherford, J. and O. Coutard. 2014. Urban energy transitions: places, processes and politics of Socio-technical change. *Urban Studies* 51(7): 1353–1377.

In introducing a special issue on urban energy transitions, this paper surveys the history of work on urban energy and characterises the recent growth of interest in how urban areas are both the site of important dynamics in energy demand, and places that have the potential to be central to low carbon transition processes.

■ Shove, E., M. Watson and N. Spurling. 2015. Conceptualizing connections: Energy demand, infrastructures and social practices. *European Journal of Social Theory* 18(3): 274–287.

This paper examines the proposition that forms of energy consumption, including those associated with automobility, are usefully understood as outcomes of interconnected patterns of social practices, including working, shopping, visiting friends and family, going to school, and so forth. Such social practices are, however, also embedded in material infrastructural arrangements – roads as well as petrol distribution networks. This means that forms of car dependence – and associated energy demand – need to be understood as emerging through the intersection of multiple infrastructural arrangements and how these are simultaneously supporting the conduct of many varied everyday practices.

REFERENCES

Baldassare, M. 1986. *Trouble in paradise: the suburban transformation in America.* New York: Columbia University Press.

Baldassare, M. 1992. Suburban communities. *Annual Review of Sociology* 18: 475–494.

Baritz, L. 1989. *The good life: the meaning of success for the American middle class.* New York: Harper and Row.

Bickerstaff, K., I. Lorenzoni, N.F. Pidgeon, W. Poortinga and P. Simmons. 2008. Reframing nuclear power in the UK energy debate: nuclear power, climate change mitigation and radioactive waste. *Public Understanding of Science* 17(2), 145–169.

Bijker, W.E. and J. Law, (eds) 1992 *Shaping technology/building society: studies in sociotechnical change.* London: MIT Press.

BBC. 2014. Compensation for Christmas power cut from distribution firm. Available online at www.bbc.co.uk/news/business-25590636.

Bradshaw, M.J. 2009. The geopolitics of global energy security. *Geography Compass*, 3(5): 1920–1937.

Castán Broto, V., S. Glendinning, E. Dewberry, C. Walsh and M. Powell. 2014. What can we learn about transitions for sustainability from infrastructure shocks? *Technological Forecasting and Social Change* 84: 186–196.

CBC. 2013. Ice storm power outages: progress 'encouraging' in Ontario. Available online at www.cbc.ca/news/canada/ice-storm-power-outages-progress-encouraging-in-ontario-1.2473933.

Chelminsky, K. 2015. The political economy of energy access and sustainable energy transitions in Indonesia, *L'Europe en Formation* 378: 146–165.

Chaudry, M., Ekins, P., Ramachandran, K., Shakoor, A., Skea, J., Strbac, G., Wang, X. and Whitaker, J. 2011. Building a resilient UK energy system. Available online at http://nora.nerc.ac.uk/id/eprint/16648/1/UKERC_energy_2050_resilience_Res_Report_2011.pdf

Clapson, M. 2003. *Suburban century: social change and urban growth in England and the United States.* Oxford: Berg.

DEMAND Centre. 2016. *Reflections on the Lancaster power cuts of December 2015.* Available online at www.demand.ac.uk/20/01/2016/reflections-on-the-lancaster-power-cuts-of-december-2015/

Edwards, P.N. 2003. Infrastructure and modernity: force, time and social organization in the history of sociotechnical systems. In *Modernity and Technology.* T.J. Misa, P. Brey and A. Feenberg (eds). Cambridge, MA: The MIT Press, pp. 185–225.

Evans, D. 2011. Thrifty, green or frugal: reflections on sustainable consumption in a changing economic climate. *Geoforum* 42: 550–557.

Fine, B. 2013. Consumption matters. *Ephemera: Theory and Politics in Organisation* 13(2): 217–248.

Goldthau, A. 2014. Rethinking the governance of energy infrastructure: scale, decentralization and polycentrism. *Energy Research and Social Science* 1: 134–140.

Gómez, M.F. and S. Silveira. 2012. Delivering off-grid electricity systems in the Brazilian Amazon, *Energy for Sustainable Development* 16(2): 155–167.

Graham, S.C. and C. McFarlane (eds) (2014) *Infrastructural lives: urban Infrastructure in context.* Abingdon: Routledge.

Graham, S. and S. Marvin. 2001. *Splintering urbanism: networked infrastructures, technological mobilities, and the urban condition.* New York: Routledge,.

The Guardian. 2014. Iraq invasion was about oil. Available online at www.theguardian.com/environment/earth-insight/2014/mar/20/iraq-war-oil-resources-energy-peak-scarcity-economy.

The Guardian. 2017a. Tesla moves beyond electric cars with new California battery farm. Available online at www.theguardian.com/sustainable-business/2017/jan/31/tesla-battery-farm-california-energy-elon-musk.

The Guardian. 2017b. Tesla moves beyond electric cars with new California battery farm. Elon Musk: I can fix South Australia power network in 100 days or it's free. Available online at www.theguardian.com/technology/2017/mar/10/elon-musk-i-can-fix-south-australia-power-network-in-100-days-or-its-free.

Harrison, C. 2013a. The historical–geographical construction of power: electricity in Eastern North Carolina. *Local Environment* 18: 469–486.

Harrison, C. 2013b. Accomplished by means which are indefensible: electric utilities, finance, and the natural barriers to accumulation. *Geoforum* 49: 173–183.

Hering, G. 2014. Apple Campus 2: the greenest building on the planet? *The Guardian.* Available online at www.theguardian.com/sustainable-business/2014/dec/07/apple-campus-2-the-greenest-building-on-the-planet

Horrocks, S. and T. Lean. 2011. *An oral history of the electricity supply industry.* London: National Life Stories / British Library.

Howell, J.P. 2011. Powering 'progress': regulation and the development of Michigan's electricity landscape. *Annals of the Association of American Geographers* 101: 962–970.

Hughes, T.P. 1983. *Networks of power: electrification in Western society, 1880–1930.* Baltimore, MD: The John Hopkins University Press.

Karlsson, B.G. 2009. Nuclear lives: uranium mining, indigenous peoples, and development in India. *Economic and Political Weekly* 44(34): 43–49.

Kunstler, J.H. 1994. *Geography of nowhere: the rise and decline of America's man-made landscape.* New York: Simon and Schuster.

Lund, J.W., L. Bjelm, G. Bloomquist and A.K. Mortensen. 2008. Characteristics, development and utilization of geothermal resources – a Nordic perspective. *Episodes* 31(1): 140–147.

Mendonça, M. 2007. *Feed-in tariffs: accelerating the deployment of renewable energy.* London: Earthscan.

Miller, L.A. 1995. Family togetherness and the suburban ideal. *Sociological Forum* 10: 393–418.

Mitchell, C. 2008. *The political economy of renewable energy.* London: Palgrave MacMillan.

Nolden, C. 2013a. *Regulating the diffusion of renewable energy technologies: interactions between community energy and the feed-in tariff in the UK*, unpublished PhD thesis, Exeter: University of Exeter.

Nolden, C. 2013b. Governing community energy – feed-in tariffs and the development of community wind energy schemes in the United Kingdom and Germany. *Energy Policy* 63: 543–552.

Nye, D.E. 1999. *Consuming power: a social history of American energies.* Cambridge, MA: MIT Press, Cambridge MA

Orkustofnun, Norges Arktiske Universitet, Energy Styrelsen, Jar-feingi, Shetland Islands Council and Greenland Innovation Centre. 2016. *North Atlantic Energy Network*, Project Report January 2016. Available online at http://os.is/gogn/Skyrslur/OS-2016/North-Atlantic-Energy-Network-Report.pdf

Osti, G. 2016. Storage and scarcity: new practices for food, energy and water. London: Routledge.

Rajan, S.C. 2006. Auto mobility and the liberal disposition. In *Against auto mobility.* S. Bohm, C. Jones, C. Land and M. Paterson (eds). Oxford: Blackwell, pp. 113–129.

Royal Academy of Engineering. 2016. *Living without electricity: one city's experience of coping with loss of power.* London: Royal Academy of Engineering. Available online at www.raeng.org.uk/publications/reports/living-without-electricity

Rutherford, J. and O. Coutard. 2014. Urban energy transitions: places, processes and politics of Socio-technical change. *Urban Studies* 51(7): 1353–1377.

Shove, E. 2003. *Comfort, cleanliness and convenience: the social organisation of normality.* Oxford: Berg.

Shove, E., M. Panzer and M. Watson. 2012. *The dynamics of social practice: everyday life and how it changes*. London: Sage.

Shove, E., G. Walker and S. Brown. 2014. Transnational transitions: the diffusion and integration of mechanical cooling. *Urban Studies* 51: 1506–1519.

Shove, E., M. Watson and N. Spurling. 2015. Conceptualizing connections: energy demand, infrastructures and social practices. *European Journal of Social Theory* 18(3): 274–287.

Star, S.L. 1999. The ethnography of infrastructure. *American Behavioral Scientist* 43(3): 377–391.

Torriti, J., 2015. *Peak energy demand and demand side response*. Abingdon and New York: Routledge.

Unruh, G. 2012. Understanding carbon lock-in. *Energy Policy* 28(12): 817–830.

Urry, J. 2004. The 'system'of automobility. *Theory, Culture & Society* 21(4–5): 25–39.

Vannini, P. and Taggart, J. 2013. Voluntary simplicity, involuntary complexities, and the pull of remove: The radical ruralities of off-grid lifestyles. *Environment and Planning A* 45: 295–311.

Van der Vleuten, Erik, and Vincent Lagendijk. 2010. Transnational infrastructure vulnerability: The historical shaping of the 2006 European "Blackout". *Energy Policy* 38 (4): 2042–2052.

Van Vliet, B.J.M., H. Chappells and E. Shove. 2005. *Infrastructures of consumption. environmental restructuring of the utility industries*. London: Earthscan.

Verdeil, É., E. Arak, H. Bolton and J. Marcum. 2015. Governing the transition to natural gas in Mediterranean metropolis: the case of Cairo, Istanbul and Sax (Tunisia). *Energy Policy* 78: 235–245.

Walker, G. 2014. Dynamics of energy demand: change, rhythm and synchronicity. *Energy Research and the Social Sciences* 1: 49–55.

Walker, G. and N. Cass. 2007. Carbon reduction, 'the public' and renewable energy: engaging with sociotechnical configurations. *Area* 39(4): 458–469.

Walker, G. and P. Devine-Wright, P. 2008. Community renewable energy: what should it mean? *Energy Policy* 36: 497–500.

Webb, J. 2014. Evaluating urban energy systems in the UK – the implications for financing heat networks. *Science and Technology Studies* 27: 47–67.

4 Geopolitical landscapes

- Understand how the movement of energy through a spatially embedded energy chain is shaped by, and in turn gives shape to, geopolitical relations.
- Appreciate the geographically differentiated character of the 'global energy dilemma' and its implications for the geopolitics of energy.
- Apply a critical perspective to the geopolitical landscapes associated with energy systems and, in the process, develop an understanding of 'energy geopolitics' that goes beyond its traditional association with international fossil fuel supply.
- Critically examine the links between energy, territory and identity.

Power shaper. Identity forger. Territory maker. Energy systems are intimately bound to the way social and political power are organised and exercised over space. In this chapter, we show how energy systems not only express power relations, but also constitute them. We use the notion of 'geopolitical landscapes' to show how networks for the recovery, transmission, distribution and consumption of energy shape, and are shaped by, power relations among different state and non-state actors. The chapter introduces three concepts – global energy dilemmas, territory and identity – to explore energy's geopolitical landscapes. These concepts extend what it means to think about the geopolitics of energy from a conventional and overly narrow association of energy geopolitics with securing the flow of fossil fuels (primarily oil) across national borders.

Traditionally, the geopolitics of energy has been seen through the lens of hydrocarbon resources, as well as the interests of nation states and large corporate bodies. It has rested upon the assumption that territorial control over resources, such as coal, oil and gas, affords states and firms political power that can be exercised both internationally and domestically. Much attention has been focused on how relations between governments and businesses are shaped by the geographical distribution of resources – particularly in terms of leading to conflicts and wars, as well as allowing for various forms of imperialist domination. One of the most prominent examples of these contingencies can be found in the Middle East, where the drivers of military conflicts and political instability have largely been attributed to energy; likewise, it has often been argued that historically, European nations used maritime power to access energy resources in order to fuel economic development. More recently, political concerns over the energy-climate nexus have started to take centre stage, particularly concerning access to previously inaccessible resources in the Arctic and Antarctic, as well as the geopolitical consequences of climate change.

In the first part of the chapter, we move away from some of these mainstream concerns to describe how the geopolitics of carbon-intensive energy systems are increasingly shaped by the responses of states and cities to climate change and economic globalisation. We adopt a global energy dilemmas perspective (Bradshaw 2013) to examine this process of change and its implications for the geopolitics of a low-carbon transition. In the second part of the chapter, we return to the relationships between energy and territory, highlighting how infrastructures for the production, transit and consumption of energy are deeply embedded in existing national and regional structures. We use several examples from the historical development of electricity networks to show how contemporary geographies of this resource are the product of vested interests, ideologies and politics, in addition to the more conventional emphasis on technical and economic factors. In the chapter's concluding section, we reflect on the politics of national identity and sovereignty that underpin hydrocarbon circulations. We draw out the global relations of power and interdependence that have shaped, and continue to shape, the distribution and trade of resources such as oil and gas. This part of the chapter seeks to show how the physical networks that transmit these fuels also act as carriers for networks of geopolitical control.

RE-THINKING GLOBAL ENERGY DILEMMAS

The essence of the global energy dilemma is simple enough: can we have affordable, equitable and secure access to energy services that are also environmentally benign (Bradshaw 2013)? This dilemma arises because global energy demand is growing and the largest part of that demand is met by fossil fuels. Importantly, however, it is not a one-size-fits-all phenomenon: global energy dilemmas take different forms as they play out in different ways across the globe. In short, geography matters in the relationship between energy consumption, energy security and climate change. For the high-energy societies of the industrialised world (the OECD), who have been responsible for most historical carbon emissions, the challenge is to rapidly decarbonise their energy systems while sustaining secure and affordable access to energy services. For emerging economies – the likes of China and India – the key challenge is

resource curse

also known as the paradox of plenty (Karl 1997), this refers to the socio-economic and political challenges that accompany, inter alia, the presence of significant energy resources in a given country. It has been argued that states that are rich in hydrocarbon resources have tended to achieve a lower level of economic development than would otherwise be expected, while being more authoritarian, conflict-prone and unstable.

Dutch disease

closely related to the resource curse are situations when a national economy is flooded by an inflow of foreign currency – often as a result of the discovery of large hydrocarbon reserves. This will normally appreciate the country's domestic currency, making its products less competitive on export markets and leading to the migration of non-extractive industries to lower-cost locations. The phenomenon is named after the Dutch economic crisis that followed the discovery of North Sea natural gas in the 1960s.

unconventional oil and gas

these are sources of oil and gas that, due to their very low permeability, require methods for extraction that are not normally necessary in conventional extraction of oil and gas. They include shale gas, shale oil (tight oil), tight sands and coal bed methane (CBM). Their extraction involves a combination of horizontal drilling and high-volume hydraulic fracturing (also known as fracking). Bituminous sands (tar sands) are also unconventional forms of oil as the source material

securing sufficient energy to meet the needs of their growing economies, while seeking an early peak to emissions and then significant reductions. For countries dependent on fossil fuel exports, the challenge is to secure development benefits from extracting hydrocarbons and avoid the **resource curse** and so-called **Dutch disease** by diversifying their economies. For other parts of the world – the global South – the challenge is to achieve universal access to modern energy services (Sustainable Development Goal 7).

The efforts of cities, states and firms to address (or avoid) these differentiated challenges will define the social and geographical distribution of the 'goods' and 'bads' associated with the provision of energy services for years to come. The result of these efforts is a dynamic geopolitical landscape of winners and losers, shaped by competitive tensions and collaborative alliances. A global energy dilemmas perspective, therefore, focuses on the underlying processes driving the geographies of energy demand and carbon emissions, and the place-specific ways in which globalisation, energy security and climate change policy interact with one another in different contexts. Its effect is to recast the geopolitics of energy as more than a zero-sum game in which governments and businesses struggle over access to oil resources and markets. A global energy dilemmas perspective highlights the disruptive influence on the geopolitical landscape of two developments in relation to high-carbon fuels. The first is the growing commitment to address the threat of climate change; and the second is the impact of the **unconventional oil and gas** revolution.

The Kyoto Protocol, an international agreement linked to the United Nations Framework Convention on Climate Change, was adopted in 1997 but only enacted in 2005. During the period to 2012, 37 industrialised countries and the European Community committed to reduce GHG emissions by an average of 5% against 1990 levels. During the second commitment period (from 2013–2022), they committed to reduce GHG emissions by at least 18% below 1990 levels. At the end of 2015, at the COP-21 meeting in Paris, a new global climate change agreement was reached (the Paris Agreement/Accord de Paris). This time a bottom-up approach was adopted with all states having the opportunity to put forward their planned contribution to "holding the increase in the global average temperature to well below 2°C above pre-industrial levels and pursuing efforts to limit the temperature increase to 1.5°C above pre-industrial levels" (United Nations, n.d.). Current commitments fall short of what is needed, but the ambition is to ratchet commitments upward through five-yearly reviews to get the world on the path needed to reduce the risks and impacts of climate change. The Paris Agreement entered force on 4 November 2016 and, at the time of writing, 127 of the 197 parties (states) had ratified the agreement, including China and the United States. President Donald Trump overturned the US commitment to the Paris Agreement and pledged to stop funding UN climate change activities; this is undoubtedly a retrograde step and one that goes against the efforts and interests of many US cities, states and businesses, but the rest of the world has renewed its commitment to taking serious action on climate change. These commitments are important because they constrain the role fossil fuels can play in the global energy system moving forward.

One way to think about that constraint is the notion of a global 'carbon budget'. We know with a high degree of certainty that fossil fuel combustion is the single largest source of GHG emissions; and, with similar certainty, we know from ice core data that the current level of carbon dioxide concentration in the atmosphere – in July 2017 it was 407.25 parts per million (ppm) – has not been higher for at least 800,000 years (and, using proxy data in lieu of ice cores, quite probably for as much as 25 million years). The consensus among climate change scientists is that 450 ppm is consistent with at least a 50% chance of limiting the average global temperature increase to 2°C, although there is an inevitable degree of uncertainty around this judgment. The International Energy Agency's (IEA) 450 scenario that constrains global warming to 2°C limits total remaining cumulative energy-related CO_2 emissions between 2015 and 2100 to 1,000 $GTCO_2$ (IEA 2014). The other side of the equation relates to the amount of fossil fuels that are available to the global energy system, under conditions where fossil fuel production itself is geographically highly uneven. It is possible to calculate the amount of GHG emissions that would be produced if we were to burn all currently known proven reserves of coal, oil and gas that are technically and commercially available. From this calculation it is clear we already have access to sufficient reserves to significantly exceed the remaining carbon budget. It is highly likely, therefore, that a sizeable amount of current fossil fuel reserves will be unburnable because we cannot exploit them and remain within the planetary carbon budget.

From a geopolitical perspective, the critical question is *whose fossil fuels will become unburnable*? This is not a straightforward matter as the rate at which we consume the remaining carbon budget depends on the global energy mix moving forward. Coal is more carbon intensive than oil, and oil is more carbon intensive than natural gas (**Chapter 1**), so failure to constrain use of coal leaves less of an opportunity to use oil and gas. The problem is that for many emerging and developing economies coal is the most secure and affordable fossil fuel. A highly influential paper in the journal *Nature* (McGlade and Ekins 2015) sought to measure and map the extent of unburnable carbon if the planet is to have a reasonable chance of avoiding a 2°C rise in global temperature. They concluded that "globally, a third of oil reserves, half of gas reserves and over 80 per cent of current coal reserves should remain unused

is not liquid and must be 'upgraded' to produce a synthetic crude. Similarly, oil shales containing kerogen (a solid mixture of organic compounds) can be mined and processed to yield a substitute for crude oil.

geo-energy space
describes the formation of a regional energy block, such as the EU: the term is used to both explain, and advocate for, the existence of common energy policies in this domain (Mañé-Estrada 2006). Proponents of the EU as a geo-energy space highlight the importance of creating regional markets for primary energy, often citing the growing importance of gas consumption within the EU as a rationale Elements that could support this strategy include expanding and strengthening the institutional framework for EU-Russia relations, and ensuring that Turkey is incorporated in EU energy policies.

stranded assets

the concept of a stranded (or impaired) asset comes from accountancy, but in the context of decarbonisation the IEA (2013: 93) defines them as "those investments which have already been made but, which, at some time prior to the end of their economic life (as assumed at the investment decision point), are no longer able to earn and economic return". An example would be a power station that is closed prematurely due to changing regulations governing its emissions that make it unprofitable. It is increasingly used to describe a possible future for many of the coal, oil and gas reserves currently held by corporations and states.

unburnable carbon

an idea that gained popular currency at the turn of this decade following the publication of an article in *Rolling Stone* by the environmentalist Bill McKibben (2011) entitled 'Global Warming's Terrifying New Math'. The article popularised the work of the London-based NGO Carbon Tracker Initiative (2011) who define unburnable carbon as "fossil fuel energy sources which cannot be burnt if the world is to adhere to a given carbon budget".

from 2010 to 2050 to meet the target of 2°C" (p. 187). Furthermore, they showed that "development of resources in the Arctic and any increase in unconventional oil production are incommensurate with efforts to limit average global warming to 2°C" (p. 187). They also demonstrated that constraints on fossil fuel consumption are even greater without carbon capture and storage technologies to reduce future emissions from power generation and industrial processes. The implication is that economically marginal, high carbon resources face an elevated risk of abandonment in a carbon-constrained world (see **stranded assets**). Efforts to account for carbon intensity, therefore, raise difficult budgetary concerns for countries dependent on the production and export of oil and coal for revenue. It is no surprise, then, that in 2016 the world's largest oil exporter, Saudi Arabia, launched a national plan (Saudi Vision 2030) to accelerate economic diversification and prepare the country for an era beyond oil. Carbon accounting also raises strategic planning and corporate valuation issues for companies and financial firms. In 2016 the international Financial Stability Board established a Task Force on Climate-related Financial Disclosures to improve climate-related financial risk disclosures by companies.

The notion of **unburnable carbon** has been championed by many environmental NGOs and think tanks. The environmental and financial risks associated with high carbon projects have propelled popular campaigns to 'keep the coal-in-the-hole, the oil-in-the-soil and the tar-sands-in-the-land' (e.g. Oilwatch), as well as the growing fossil fuel divestment movement that seeks to prevent the further proliferation of high-carbon resource landscapes (Figure 4.1). For example, the think tank Carbon Tracker examines the impact of climate change on financial markets and argues that companies, fund managers and banks – both private and state – need to pay much greater attention to the impact of climate change policy on the future value and performance of fossil fuel companies (see **Example 1**). Oil Change International (2016) is another example of an organisation that aims to highlight the real costs of fossil fuels and promote the transition to a low carbon energy future. From their perspective, the urgency of responding to climate change is so great that "no new fossil fuel extraction or transportation infrastructure should be built, and governments should grant no new permits for them". Even the International Oil Company (IOC) BP is prepared to admit in its *Energy Outlook to 2035* (2016: 15) that "there is an abundance of oil resources" and that "under most scenarios cumulative oil demand amounts to less than half of today's technically recoverable reserves". However, that does not mean that we should stop investing in oil and gas, as suggested by the divestment campaign. As the IEA (2016) warns in its latest *World Energy Outlook,* even in their 450 scenario (i.e. a situation that constraints global warming to 2°C), fossil fuels still account for 58% of global primary energy consumption in

Figure 4.1
Keep the oil in
the soil and
the coal in the
hole: a poster
expressing
resistance to
fossil fuel
extraction, from
COP 17 (held in
Durban, South
Africa as part
of negotiations
around the
UNFCCC)
(photo credit:
WWF
LAI/Denise
Oliveira)

EXAMPLE 1: FOSSIL FUEL DIVESTMENT

The concept of divestment is not new and previous divestments campaigns have sought to protest the Apartheid regime in South Africa or discourage investment in the tobacco industry, for example. Fossil fuel divestment campaigns date from the 1990s, and the most recent incarnation encourages investors, and particularly large institutional investors such as sovereign wealth funds, university endowments and pensions funds, to give up their holdings of fossil fuels stocks and reallocate those funds to climate-friendly alternatives. Ayling and Gunningham (2017) have recently charted the evolution of this 'transnational advocacy network'; while Healy and Debski (2016) have explored the student-led fossil divestment campaign in US higher education. Advocates of divestment appeal to a combination of ethical and self-interested arguments: in the context of decarbonisation and slowing demand growth, financial holdings (e.g. shares) in fossil fuel companies may not perform well. They evoke concepts such as unburnable carbon and standed assets to suggest that the future value of fossil fuel companies will be compromised. The concept is gaining traction as central banks, private banks and the insurance industry recognise climate change policy as a source of future financial risk. Not surprisingly, international oil companies have tended to respond by assuring investors that it will not be their assets that get stranded, though some of them now admit that not all current proven reserves will be burnt. Check out the status of the divestment campaign on your university's campus.

2040, down from 81% in 2014. The key issue in the context of carbon constraints is that fossil fuel demand growth must slow, peak and decline in the coming decades if we are to avoid "catastrophic climate change", the term used in climate change science.

This means that the traditional conception of the geopolitics of energy security as a competition between fossil fuel producing and consuming states in a world of scarcity and ever-increasing demand, is not the future that we are likely to face. Increasingly, security of demand, not security of supply, will be the greatest challenge for fossil fuel companies and reserve holding states. Equally, the future energy trajectories of emerging and developing regions will not mimic the history of the industrialised economies; rather, these states will shape their own trajectories and energy landscapes within the wider context of the low carbon energy transition.

So, how have we arrived at this world of fossil fuel abundance? The main reason has been the so-called unconventional fossil fuel revolution in North America. Geologists have known for a long time that shale rock contains hydrocarbons: indeed, shale rocks are the source of conventional oil and gas deposits that have accumulated in more permeable rock formations that allow liquid and gases to pass through them. However, the challenge has been how to free the trapped oil and gas molecules from impermeable shale rock. The answer lies in the combination of two techniques that have a long history in the oil and gas industry: horizontal drilling and hydraulic fracturing. Over a thirty-year period, a technique was perfected that uses horizontal drilling to access the shale layers and then pumps high volumes of water at high pressure into the rock to create fractures (hence fracking). These fractures are then held open by sand grains, or 'propants', and this enables the gas or oil to follow into the well and to the surface. The use of 3D seismic surveys and complex modelling has also helped drillers to understand better the shale formations. But the shale gas industry is different from conventional oil and gas as the production from each well rapidly peaks and declines, resulting in a short period of relatively high production and a much longer period of relatively low production. This means that to maintain levels of production it is necessary to keep drilling new wells. If the number of wells drilled – the rig count – falls, so production subsequently declines. Although the cost of drilling an individual well is modest compared with a conventional gas field, this need to constantly drill new wells is costly and means that unconventional oil and gas is vulnerable to downturns in price. The intrusive and cumulative nature of unconventional oil and gas development is not without controversy and many environmental groups and affected communities maintain that it is environmentally unsound and has a negative impact on human health. Protestors also argue that there is no need to develop new unconventional resources as we already have access to more conventional resources that we can burn.

Over the last decade, the development of shale gas and tight oil has dramatically changed the energy situation in the United States, with global implications. According to the US Energy Information Administration (EIA), petroleum imports into the US have declined since total consumption peaked in 2005; but the level of domestic oil production has increased significantly since 2010 thanks to tight oil production. Consequently, the level of US oil import dependence has fallen significantly, although it still imported 24% of its petroleum needs in 2015. The situation in natural gas is even more dramatic. At the turn of the century the US was preparing to become a significant importer of liquefied natural gas (LNG); but in early 2016 the first LNG export terminal started operations, with more planned. By the mid-2020s the US is predicted to become a net energy exporter. Already, this new-found energy abundance has changed market expectations and OPEC has had to take drastic action to reclaim its market share. With a new President in the Whitehouse determined to "Put America first" and who is also sceptical about climate change, it is uncertain what the geopolitical consequences of US energy independence might be. Equally, the geopolitical status of OPEC is challenged by America's growing energy independence in ways that are not yet certain.

So far, North America's unconventional oil and gas revolution has struggled to travel, though there are significant shale resources worldwide – particularly in China, Russia and Argentina. In some cases, the new techniques have not worked with local geology, in others there is significant public resistance to shale development. Nonetheless, unconventional oil and gas has already been global game changer. Thus, we can conclude that we have two contradictory tendencies – carbon constraints and fossil fuel abundance – that require that we rethink the geopolitics of global energy security.

Consequences and challenges: geopolitics of the low carbon energy transition

Elsewhere (**Chapters 9 and 10**) we explore the notion of energy transitions and the need to orchestrate a transition away from the current 'high carbon' fossil fuel energy system to a future 'low carbon' system. At present, the low carbon system comprises large-scale hydroelectricity, new renewables (mainly wind and solar) and nuclear power. There is considerable debate about how long it will take to bring about this transition. The record of past energy transitions (**Chapter 9**) tells us that it is a matter of thirty to forty years as a minimum, but the rapid growth of renewable energy – albeit from a low level – suggests that it could happen more quickly than expected. However, it will not be without consequences and there will be winners and losers on both sides of the transition.

The global energy dilemmas framework suggests that we face a triple challenge: first, to accelerate improvements in energy intensity, that is to reduce the amount of energy used per unit of economic output (energy efficiency and demand reduction, see **Chapter 2**); second, to reduce the carbon intensity of energy use, that is to reduce the amount of CO_2 produced per unit of energy used (decarbonsiation); and to achieve the above in ways that are secure, affordable and equitable (and that does not threaten economic growth). While some would claim that the very nature of economic progress needs to change, it is clear that the relationship between energy and economy must fundamentally change, as must the ways in which we secure access to energy services. The process of energy transition clearly challenges the incumbent fossil fuel system, which is the most obvious loser in the long-term, and creates the potential for a new set of winners in the coming low carbon system (see **Example 2**). Consequently, a low-carbon energy transition will fundamentally change the geopolitics of global energy security.

To date, relatively little thinking has gone into the geopolitical consequences of a low carbon transition and the new geopolitical landscape that are emerging. There are the beginnings of a literature on the geopolitical characteristics of renewable energy, however, and here we extend a recent paper by Scholten and Bosman (2016). First, renewable energy resources are neither as scarce nor geographically constrained as fossil fuels, although there is significant geographical variation in the effectiveness of wind and solar power and the resultant geographies of production are quite different from those of oil and gas. Second,

EXAMPLE 2: THE GEOPOLITICS OF RENEWABLE ENERGY

As the global energy supply mix moves towards a greater reliance on renewables and electricity, it is anticipated that existing geopolitical and geoeconomic landscapes will undergo several seismic shifts. For one, the foreign policy influence of oil-exporting states is likely to decrease, with major 'emerging' hydrocarbon resource-poor powers – such as China, India and Turkey – playing a more important role in the global energy market. At the same time, the production and consumption of renewable energy is gaining growing political power, both within states and in terms of international relations. In countries like Germany, where low carbon transition has been a major national project, investment in renewable infrastructure is a key factor in shaping domestic politics. Transition towards renewable, low-carbon energy is has become a driver of regional co-operation among states, associated with electricity network and energy storage requirements: electricity will become increasingly important as a primary means of cross-border energy trade, occupying the role of today's oil and gas pipelines and ocean tankers (Forbes 2018). There are other geopolitical issues too, such as the supply chain for rare earth minerals and other materials used in the renewables sector (see **Chapter 1**); the politics of knowledge control around patents and technology, given that renewable energy development hinges upon technical innovation; and the differential capacity of states to move towards infrastructures and practices of energy demand more amenable to renewable energy supply. The geopolitics of renewable energy may, therefore, underpin the emergence of a new multipolar world order, in which states and other sites (e.g. city-regions) that have a strong investment record in these types of resources will take centre stage. The International Renewable Energy Agency (IRENA) launched a Global Commission on the Geopolitics of Energy Transformation in 2018 to better understand the short and long-term impact of renewables on the geopolitical landscape (Forbes 2018).

many renewable sources, including solar and wind, are intermittent. This requires sufficient back-up capacity to cover periods of low generation and/or significant improvements in the cost and efficiency of electricity storage. Increased interconnection is also a means of managing intermittency, but this requires inter-state cooperation and access to surplus capacity when needed. Third, renewable electricity generation technologies can be much smaller than conventional technologies, though large-scale offshore wind farms are an exception, and this suggests the development of a more distributed energy system whereby every land- or even roof owner is a potential electricity producer. The wider consequence is that the low carbon transition is bringing with it a re-scaling of national energy systems (Bridge et al. 2013). Fourth, renewable energy generation (and storage) requires access to new resources, such as lithium and rare earth metals, which is creating new geopolitics around their production and trade (**Chapter 1**). At the same time, the countries and companies that develop and control the large-scale production of low carbon generation and storage technologies are likely to gain a significant advantage in what many now call the 'green economy'. Finally, electricity can be expected to become the dominant energy carrier in a world dominated by renewable energy. Already the power sector is commanding a larger share of global energy investment than oil and gas and this trend is set to accelerate. However, it should not be assumed that universal access to modern energy services will delivered via the roll out of national transmission systems; instead there will be a combination of grid-based systems, regional and mini-grids and even community level solutions using a variety of different technologies. The result will be much more varied energy landscapes than the centralised grid-based systems that dominate the industrial economies.

Transition to a low carbon energy system will not happen overnight and it will be a matter of decades rather than years (Sovacool 2016). But the pace of change is accelerating, and those who have benefitted the most from the incumbent fossil fuel system are finally beginning to realise their days are numbered. The CEO of Shell recently talked of preparing his company for a world where oil prices are 'lower forever'. As noted earlier, the process of transition itself is likely to be particularly challenging as winners and losers jockey for position in an increasingly volatile and uncertain energy world. A final point to recognise is that a complex situation will be made even more challenging if we fail to constrain GHG emissions and must confront the consequences of catastrophic climate change.

CONNECTING ENERGY AND TERRITORY

This section examines the relationship between territory and energy, and presents an alternative reading of energy landscapes to the traditional geopolitical focus on the international supply of fossil fuels. The notion of 'territory' is frequently used in a range of social science disciplines to describe the political organisation of space. In human geography and political science, for example, territory describes the exercise, ordering and governance of power relations across space. Territory is an expression of the desire of a social, economic or political actor (e.g. a state or corporation) to extend its jurisdiction over a given physical area. The territorial practices associated with political or commercial power are often underpinned by the infrastructural systems for capturing, moving and consuming energy. The extension of energy dependencies beyond the confines of conventional state borders – via transnational energy resource flows and cross-border investment in energy infrastructures – has destabilised the notion that state territories neatly conform to national boundaries. Thus, the concept of territory is increasingly used by social scientists to refer to the outcome of networked socio-technical processes that may extend across a wide range of scales and spaces, rather than simply the area enclosed by the formal boundaries of a state.

energy territory

territory is normally understood as a physical area that falls under a particular political jurisdiction. Energy networks involve supply chains that extend across large geographical spaces, while being connected by shared regulatory and political standards. Thus, in the case of energy, territory can be seen as the outcome of networked socio-technical practices. This means that energy territories are not contiguous with existing urban, regional or national boundaries but rather, are associated with relations of power that connect consumers and suppliers across a multitude of physically disparate spatial realms.

The co-constitutive relationship between energy and territory (i.e. how each shapes the other) is illuminated by studies of large socio-technical systems, which highlight the processes involved in the construction and functioning of territorial infrastructure networks (**Chapter 3**). Research shows that **energy territories** are frequently constructed in an incremental manner, and are not easily created 'from above': historically, energy networks – such as pipelines and electricity grids – have expanded from the bottom up, involving the growth of multiple installations via stepwise economic investment. This has been accompanied by wider political consolidation of the urban, regional or national spaces surrounding these networks: in this way, energy systems can have the effect of constituting political relations and bringing new territories into being. Building new energy infrastructures and getting them to work also requires the formulation of technical standards and conventions of practice. Increasingly, this has involved inter-state cooperation and the implementation of shared infrastructural norms over distinct territorial spaces. Common examples include differences in voltage levels and electric plug connectors, and variations in the calorific (i.e. heat) content of gas sent through transmission and distribution systems. In this way energy infrastructures are frequently 'learned as part of membership' within a broader territorial structure, highlighting the identity-building role that energy infrastructure systems often play. The creation of single electricity and gas markets in the EU is a contemporary example of this process.

The creation of national electricity and gas grids in Europe and North America – arguably one of the "greatest engineering achievements of the twentieth century" (Constable and Somerville 2003) – took place gradually over several decades. It involved the transformation of small-scale, municipal systems into national networks under the influence of wider political and economic dynamics. Security of supply concerns that emerged during World War I had a decisive influence on the development of central state policies for the creation of unified power systems in the United States and the United Kingdom, for example. In the US, electric utilities joined forces after the 1920s to increase the efficiency and stability of their operations, so that electrical networks were subsequently recognised as public goods at the national scale by the 1934 Public Utility Holding Company Act (**Chapter 2**). Key events in the electrification history of the UK involve the standardisation and synchronisation of power networks by the Central Electricity Board in 1926, and establishment of the National Grid in 1938. The territorial form in which standardised energy networks initially developed can create strong path dependencies. Japan's electricity grid, for example, is divided into two frequency zones – a situation that creates all kinds of technical and policy problems – as a result of historical legacies. Eastern Japan follows the German model and runs at 50 Hz, while the western part of Japan uses the American frequency of 60 Hz. This can be traced back to the first purchases of generators for Tokyo in 1895 (from the German company AEG) and Osaka in 1896 (from US-based General Electric).

Initial electricity network development in much of South America, Africa and Asia was an outcome of colonial objectives, combined with local political and cultural circumstances. The development of India's power transmission system during the colonial Madras presidency – in place between 1900 and 1947 – was primarily influenced by regional politics based on linguistic differences, although more general political ideological battles also played a role. It led to a major economic divide between northern and southern parts of the country, while fuelling regional, caste and communal tensions (Rao 2010). There is limited research on the history of African electrification, but research on the electrification of Bulawayo in colonial Zimbabwe (Rhodesia) from 1894 to 1939 has shown how the colonial politics of electrical power played a major role in deepening existing racial and social divisions. Electricity was harnessed "to control and police the underprivileged" via a "parochial project that benefited white settler residents of the town almost to the total exclusion of Africans" (Chikowero 2007: 207). Electric power development also played a significant role in generating, strengthening and signifying socio-economic and political divisions between Arabs and Jews during British rule of Palestine in the 1920s. Colonial interests supported the Jewish–Zionist dominated Jaffa Electric Company in the initial construction of a grid that served the towns of Jaffa and Tel Aviv, as well as nearby predominantly Jewish settlements and British military installations. Electrification thereby created a physical infrastructure of separation along ethnic and spatial lines, well before the formal establishment of partition plans.

These examples indicate how the territorial diffusion and re-scaling of energy infrastructures is driven by more than technical or economic considerations. Energy has also been integral to the making of political projects in the broadest sense, and at a range of different geographical scales. These include the creation of the nation state and the making of the modern citizen. Social science research on the development and expansion of electric power systems in the early twentieth century has shown how they followed four evolutionary phases encapsulated, briefly, as 'invention and development', 'technology transfer', 'system growth' and 'substantial momentum' (Hughes 1993). However, progression from one stage to the next has never been linear and straightforward, with the emergence of various setbacks, difficulties and digressions being the rule rather than the exception. More importantly, political and cultural factors have been shown to play a potentially more important role than traditional concerns over technical efficiency, economic rationality, and cost-effectiveness. In countries like the United States, Germany and the United Kingdom, decisions over how and where power networks would be constructed were largely made by the managers of utility companies, rather than the inventors and engineers whose work made the systems technically feasible in the first place. The evolution of electrical networks in London, for example, was characterised by frequent political conflicts and stalemates among the interests of local government authorities, ideologies of 'municipal socialism', and the work of private companies. At the same time, the relations between political and technological power that characterised electrical network development in Berlin were more closely integrated, thanks to the presence of practices of co-ordination and co-operation.

Even though the American model of power sector development largely relied on the initiative of commercial and corporate actors, the manner in which electricity networks expanded in the country was frequently neither the most technically nor economically efficient. Electrification in the United States was driven by a relatively narrow set of powerful economic actors, who achieved their objectives by mobilising "shared personal understandings, social connections, organizational conditions and historical opportunities" that were available to them (Granovetter and McGuire 1998: 149). Particularly important was the ability of these economic actors to influence the creation, via informal political ties, of a high degree of organisational and legal uniformity in electricity markets. The effect was a "suppression of diversity" that weakened the technological and institutional adaptability of the industry for decades. The expansion of electricity networks also had significant cultural overtones: applications of electrical power (such as illumination) imbued urban and regional landscapes with 'transcendent significance' and propagated a national enthusiasm for, and faith in, the transformative power of electricity and technological progress – creating a national aesthetic that cultural historian David Nye terms the 'technological sublime' (Nye 1996). The electrification of large cities provided a method for transforming the social uses and meanings of public space, via spectacular displays of modernity in theatres, fairs, tall buildings, streets and squares (see **Chapter 9**). But the transfer of such practices and technologies beyond large urban centres was anything but smooth: acts of 'transformative resistance' among rural citizens meant that the introduction of new technologies was both opposed and modified, via the use of electric lights and appliances in ways different to those prescribed by the authorities (Kline 2002).

POLITICS, NATIONAL IDENTITY AND ENERGY

Gas systems also offer a good example of the embeddedness of energy infra-structures in broader dynamics of political power, identity and belonging (see **Case Study**). Present systems for the distribution of natural gas draw their roots from town gas networks developed during the nineteenth century. These networks distributed gas that was manufactured from coal, with the gas used primarily for street lighting (see **Chapter 9**). They were eventually replaced by the electricity networks described above reflecting, in part, the perceived superior environmental and aesthetic performance of electricity over gas. However, the discovery of conventional gas fields as a result of the pursuit of oil led to far-reaching changes in the energy chain of this resource (Figure 4.2). Notable finds included the Hugoton reservoir in Kansas (1922), the Groningen field in the Netherlands (1959), and the Urengoy gas field in West Siberia (1966). Thanks to innovation and national-scale investment in gas transportation, these newly found supply sources were connected with rapidly increasing numbers of consumers in their host countries, leading to changes in the fuel mix away from coal and oil. Conventional gas growth saw a further expansion following the oil shocks of 1973 and 1978, which prompted governments and utilities in Western Europe and America to develop active policies for the diversification of energy supply. This also intensified international trade via pipelines and LNG, which started to play a more prominent role. Adding a further layer of complexity to an already multi-layered international energy landscape has been the expansion of unconventional gas exploitation in the

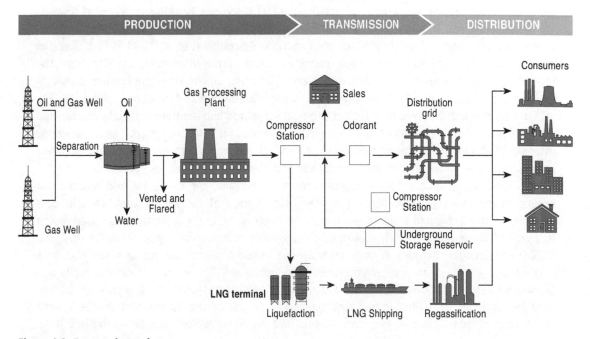

Figure 4.2 Energy chains for gas

CASE STUDY: GEOPOLITICAL LANDSCAPES OF GAS TRANSIT IN EUROPE

Questions of political power, national sovereignty and international relations have played a prominent role in concerns over gas supply and transit. These geopolitical concerns are amplified by the material nature of gas pipelines, which establish a physical link between energy producers and consumers, often across vast geographical distances. An emblematic example is provided by the European Union, whose economic and political relations have been influenced by dependence on gas imports from neighbouring regions. The threat of supply interruptions has been one of the main driving forces behind the establishment of common EU policies in the energy domain – a complex and difficult project that has been often hindered by the conflicting national interests of various EU energy states. Recently, the formulation of a new EU energy policy has taken been taken further via the adoption of the Energy Union Strategic Framework, which promises to bring about greater energy security via the integration and liberalisation of energy markets.

The ability of energy-related geopolitical tensions to polarise public opinion became evident during the construction of the Nord Stream gas pipeline, which provided a direct undersea link from Russia to Germany but was widely opposed in public and media debates within neighbouring states that were bypassed by the project, including Poland, Lithuania, Latvia and Estonia (Bouzarovski and Konieczny 2010). Experts have suggested the creation of a common **geo-energy space** as one of the ways of overcoming such tensions in the future. This would allow for the integrated planning of sites of energy recovery, demand and transit, ensuring that resource extraction takes place in suitable locations and is accompanied by smooth network connections.

EU attempts to develop collective policies for energy security are particularly evident in the domain of gas transit, where collective financial resources have been deployed towards the construction of new intra-country interconnections as a means of generating flexibility, interoperability and redundancy in an open market. Bouzarovski et al. (2015) find that this geopolitical and regulatory landscape operates across multiple sectors, starting from the formulation of official policy frameworks and including day-to-day functioning of socio-technical infrastructures such as pipelines, pumping stations and storage facilities. A key role in this context is played by different physical standards and regulatory devices established at multi-lateral level (see **Chapter 3**). These includes codes for balancing transmission networks, inter-border capacity allocation mechanisms and 'congestion management' procedures, inter-operability rules between different countries, and framework guidelines for tariffs.

The European Union's Southern gas corridor exemplifies the character and direction of recent energy security thinking within the bloc. Aimed at increasing the "security and diversity of the European Union's gas supply by bringing natural gas from the Caspian region to Europe" (www.tap-ag.com/), it is one of the world's most complex gas transit chains. The 3,500 km corridor includes several separate but linked projects, traversing seven countries and engaging more than a dozen major energy companies. These include drilling wells and producing gas in the Caspian Sea, investment in natural gas processing plants, and three pipeline projects stretching from Azerbaijan and Georgia, across Turkey and the Bosphorus strait, onto northern Greece, northern Albania and the Adriatic Sea, and extending into Italy.

Further investment in the Italian grid is expected to allow additional interconnections with Central and Western Europe, and a number of countries in the western Balkans – some of which do not have any piped gas networks – are also expected to be able to join. To a certain extent, this complex multilateral initiative replaces and supplants the previously unsuccessful and less ambitious Nabucco pipeline, which was criticised for lacking a sound commercial basis and political support.

United States, with its associated technological, economic, environmental and political impacts.

While gas is a global energy resource today, its economic geography is largely fragmented into three macro-regions: Europe (characterised by the heavy reliance on overland import pipelines), North America (where LNG is playing an increasing role) and East Asia (where the distinguishing factor is the predominance of LNG). Increasing trade links within the three spaces are driven mainly by the globalisation of LNG, where the second (Qatar) and third (US) largest exporting countries trade across all regions, although the largest exporter (Australia) is mainly focused on the Asian market. Russia, which falls into fourth place in this ranking is also closely connected with Asian LNG demand, while Europe provides the main export for the fifth largest exporter of this resource – Algeria (see Bridge and Bradshaw (2017) for a geographical analysis of the LNG industry). The existence of several distinct modes of gas provision points to the importance of national scale institutional and political arrangements in shaping the geography of natural gas. Indeed, it has been shown that the way countries choose to regulate and co-ordinate their gas sector has a significant impact on the physical and territorial nature of the resulting infrastructural network. Countries with gas systems at low to intermediate levels of development (including most continental European states as well as major Asian states such as China and Iran) tend to possess gas systems that are nationally anchored on large-scale supply and demand sources in a hub-and-spoke manner. A greater level of economic and organisational advancement – such as that seen in the US, the Netherlands, Canada, the UK and Russia – is associated with gas networks that extend well beyond national boundaries and involve overlapping modes of supply, transit and demand. The territorial reach of gas trading hubs in Europe, for example, does not strictly follow national borders.

Hydrocarbons have also played a role in shaping wider questions of national identity and security. This is particularly evident in the case of oil: as part of a broader trend that developed during twentieth century, state sovereignty over oil resources has been seen as vital for a range of economic, financial, environmental and resource reasons (Bridge and Le Billon 2017). The suggestion that nation states should be able to control the oil resources located within their own territories gained international prominence after World War II, having become embedded in post-colonial struggles for independence and autonomy.

petro-state

an oil-rich state that relies on hydrocarbons for generating export revenues and national income: by this definition, Angola, Azerbaijan, Nigeria, Norway and Saudi Arabia are examples of petro-states, as is Alberta in Canada. It is frequently argued that petro-states are characterised by weak institutions, corruption and internal economic inequality, although examples like Norway indicate this need not be always the case. The geopolitical influence of petro-states may be on the wane, given lower oil prices and possibility of peak demand within a few decades.

petronationalism

a process whereby nation states aim to take government control over hydrocarbon resources, primarily via the establishment of national oil and gas companies. The process is driven by increases in energy prices, technological advances as well as political resistance towards liberalisation policies. Aside from increased state participation in the oil and gas industry, the dynamic is underpinned by tax rises, contract revisions and expropriation of assets.

This was particularly true in **petro-states** such as Mexico, Venezuela, Iran and Algeria, where oil resources were traditionally controlled by colonial powers. Nowadays, the notion of **petronationalism** is often invoked as a way of describing efforts of particular states to limit the supply chain of oil within the territorial or political confines of the nation state. Among other dynamics, petronationalism has involved the greater participation of nation state institutions in managing hydrocarbon industries via a range of fiscal, property and economic development policies. It is underpinned by the increasingly important role of national oil companies (NOCs) in the global energy market: within the top 50 companies by volume of production, over half are majority state-owned, with the top ten reserve holders – accounting for 80% of oil reserves – being represented by NOCs based outside the OECD. The increasing ability of NOCs to both shape the global energy industry and seek investment opportunities across the world has prompted concerns over the levels of economic transparency and political control associated with such entities (Smil 2003). However, there are significant differences between strategies pursued by different NOCs, especially when it comes to the distinction between companies based in oil-exporting and oil-importing states.

The political identities that are created in and through the circulation of oil extend beyond state activities to encompass sites of energy resource extraction, transit and consumption. It has been documented, for instance, that large-scale societal transformations connected to the oil industry incur multiple shifts in day-to-day human practices and relationships, while influencing the formation and articulation of cultural identities (Shever 2012). In this regard, hydrocarbon flows are experienced differently by different groups: these may include oil workers and their families, the residents of settlements that border oil-related infrastructures, and the employees of transnational corporations that direct this resource's circulation from a distance. In cases where local populations are affected by the work of the latter, local coalitions and networks may invoke, produce and articulate distinct collective identities, which in turn directly influence the local politics of the places in which they live (Valdivia 2007). Moving onto sites of consumption, a wide sociological literature – otherwise relatively poorly connected with the field of energy studies – focuses on the symbolic political meanings that are generated and transmitted through the consumption of energy. A potent example is provided by automobile ownership, which has gradually come to embody a wide range of personal identities linked to wider socio-political developments (e.g. the meanings conveyed by electric cars, or Sports Utility Vehicles). As pointed out by Huber (2009: 465), gasoline has come to fuel "broader imaginaries of a national 'American way of life'".

The geopolitics of energy are also visible in the case of nuclear power, which illustrates how 'mega energy ideas', requiring centralised, national-scale, and corporate-led control networks have been advanced in relation to geopolitical landscapes (Hecht 1998). Jasanoff and Kim (2009: 120) have argued that the mobilisation of **sociotechnical imaginaries** in the context of nuclear power development and opposition in Korea and the United States confers "the power to influence technological design, channel public expenditures, and justify the inclusion or exclusion of citizens with respect to the benefits of technological progress". A similar claim has been advanced in relation to large-scale hydropower dams, which, to cite Byrne and Toly (2006: 1), represent an 'attempt at a techno-fix of the democratic-authoritarian variety'. These examples show how large-scale projects aimed at harnessing natural resources offer the incorporation of geopolitical identities, narratives and meanings in the articulation of energy technologies.

sociotechnical imaginaries a term coined by Jasanoff and Kim (2009), highlighting the manner in which conceptions of future scientific, economic and technological growth are associated with tacit ideological and political understandings of common purpose. Somewhat linked to this is the notion of the *technological sublime,* which refers to the way large infrastructural projects – many of which are in the energy sector – represent a type of civil religion that inspires transcendental feelings while shaping the collective national character (Nye 1996).

CONCLUSION

This chapter has highlighted the variegated geopolitical landscapes of energy that have underpinned forms of energy resource capture and circulation in the past and present. We have illustrated some of the complex territorial formations associated with the extension of energy networks, and how these territories and networks constitute each other. The global energy landscape is changing, driven by the four imperatives outlined in the book's **Introduction**. New geopolitical landscapes are emerging in the process, combining established territorial structures and divisions between energy-rich and energy-poor states with new territories of energy production, transit and demand. In the coming decades, the global energy system needs to further transformed to address the challenge of climate change and extend access to modern energy services. The energy systems that respond to this challenge will constitute new territories and identities that shape the geopolitical landscapes of a future low carbon energy world.

QUESTIONS FOR DISCUSSION

- What are the potential geopolitical consequences of a world of lower oil prices?

- What new geopolitical challenges might be associated with the development of renewable energy technologies?

- Why and how are energy flows embedded in the production of territory?

- How have political interests shaped the expansion of electricity in the past?

- What is the role of national identity and sovereignty concerns in influencing oil and gas policies?

ACTIVITIES/POTENTIAL RESEARCH PROJECTS

- Consider the arguments for and against fossil fuel divestment.

- Explore the implications of thinking about Europe as a geo-energy space for UK energy policy.

RECOMMENDED READINGS

■ Bouzarovski, S., M. Bradshaw and A. Wochnik. 2015. Making territory through infrastructure: the governance of natural gas transit in Europe. *Geoforum* 64: 217–228.

This paper articulates how the geopolitical landscapes of energy territories are created and operationalised in material and institutional terms – via the regulation and functioning of transnational natural gas circulations. Using network analysis methods, it reveals the existence of overlapping geographical networks implicated in securing energy under the guidance of different actors.

■ Mitchell, T. 2009. Carbon democracy. *Economy and Society* 38(3): 399–432.

This wide-ranging article explores the political landscapes and 'democratic machineries' that took shape in the twentieth century around coal and oil. The author argues that coal opened up new democratic possibilities in industrial societies via the control it afforded mine, rail and port workers (and others) over extraordinary concentrations of energy. As energy production shifted to oil in the Middle East, however, these democratic possibilities changed: opportunities for the kind of political mobilisation central to industrial democracies weakened while, at the same time, democratic politics became focussed on the future as an unlimited horizon of growth. The article's socio-technical understanding of the links between energy and politics casts new light on the political landscapes associated with energy transitions.

■ Power, M., P. Newell, L. Baker, H. Bulkeley, J. Kirshner and A. Smith. 2016. The political economy of energy transitions in Mozambique and South Africa: the role of the rising powers. *Energy Research & Social Science*, 17: 10–19.

The paper starts from the need to develop novel theoretical understandings of South–South co-operation around renewable energy technologies alongside efforts to enhance access to hydrocarbon resources like coal and gas. It proposes a framework that attends to questions of geopolitics, diplomacy and international relations. The authors focus on the power, capacity and autonomy that states have to secure and negotiate different outcomes in the energy sector.

■ Van de Graaf, T. and A. Verbruggen. 2015. The oil endgame: strategies of oil exporters in a carbon-constrained world. *Environmental Science & Policy* 54: 456–462.

This paper explores the notion of peak oil demand and considers the threat that this poses to the future prosperity and geopolitical significance of the petro states whose GDP is largely dependent on revenue from oil exports. The articles suggest five possible strategies that such states could pursue in a future carbon-constrained world. The paper suggests that considerations of the low carbon transition need to incorporate more explicitly the behaviour of oil exporting states faced with demand destruction.

REFERENCES

Ayling, J. and N. Gunningham. 2017. Non-state governance and climate policy: the fossil fuel divestment movement. *Climate Policy* 17: 131–149.

Bouzarovski, S., M. Bradshaw and A. Wochnik. 2015. Making territory through infrastructure: the governance of natural gas transit in Europe. *Geoforum* 64: 217–228.

Bouzarovski, S. and M. Konieczny. 2010. Landscapes of paradox: public discourses and state policies in Poland's relationship with the Nord Stream pipeline. *Geopolitics* 15: 1–21.

Bradshaw, M.J. 2013. *Global energy dilemmas*. Cambridge: Polity Press.

Bridge, G., Bouzarovski, S., Bradshaw, M. and Eyre, N., 2013. Geographies of energy transition: space, place and the low-carbon economy. *Energy Policy*, 53: 331–340.

Bridge, G. and M. Bradshaw. 2017. Making a global gas market: territoriality and production networks in liquefied natural gas. *Economic Geography* 93: 215–240.

Bridge, G. and P. Le. Billon. 2017. *Oil*. London: John Wiley & Sons.

Byrne, J. and N. Toly. 2006. Energy as a social project: recovering a discourse. In *Transforming power: energy, environment, and society in conflict*. J. Byrne, N. Toly and L. Glover (eds). New Brunswick: Transactions, pp. 1–34.

Carbon Tracker Initiative. 2011. Unburnable carbon – are the world's financial markets carrying a carbon bubble? London: Carbon Tracker Initiative. Available online at www.carbontracker.org/wp-content/uploads/2014/09/Unburnable-Carbon-Full-rev2-1.pdf

Chikowero, M. 2007. Subalternating currents: electrification and power politics in Bulawayo, Colonial Zimbabwe, 1894–1939. *Journal of Southern African Studies* 33: 287–306.

Constable, G, and B. Somerville 2003. *A century of innovation: Twenty engineering achievements that transformed our lives*. Washington D.C., National Academy of Engineering.

Forbes (2018) Renewables industry starts to grapple with the geopolitical impact of shift to green energy. January 14 2018. Available online at www.forbes.com/sites/dominicdudley/2018/01/14/renewable-energy-geopolitical-impact/

Granovetter, M. and P. McGuire. 1998. The making of an industry: electricity in the United States. *The Sociological Review* 46: 147–173.

Healy, N. and J. Debski. 2016. Fossil fuel divestment: implications for the future of sustainability discourse and action within higher education. *Local Environment*. DOI: 10.1080/13549839.2016.1256382

Hecht, G. 1998. *The radiance of France: nuclear power and national identity after World War II.* Cambridge, MA: The MIT Press.

Huber, M.T. 2009. The use of gasoline: value, oil, and the 'American way of life'. *Antipode* 41: 465–486.

Hughes, T.P. 1993. *Networks of power: electrification in western society, 1880–1930.* Baltimore, MD: Johns Hopkins University Press.

IEA 2013. Redrawing the energy climate map, World Energy Outlook Special Report. Available online at www.worldenergyoutlook.org/media/weowebsite/2013/energy climatemap/RedrawingEnergyClimateMap.pdf

IEA 2014. *Energy technology perspectives 2014.* Paris: International Energy Agency.

IEA 2016. *The world energy outlook 2016.* Paris: International Energy Agency

Jasanoff, S. and S.-H. Kim. 2009. Containing the atom: sociotechnical imaginaries and nuclear power in the United States and South Korea. *Minerva* 47: 119–146.

Karl, T. 1997. *The paradox of plenty: oil booms and petro-states.* Berkeley, CA: University of California Press.

Kline, R.R. 2002. Resisting development, reinventing modernity: rural electrification in the United States before World War II. *Environmental Values* 11: 327–344.

Mañé-Estrada, A. 2006. European energy security: towards the creation of the geo-energy space. *Energy Policy* 34: 3773–3786.

McGlade, C. and P. Ekins. 2015. The geographical distribution of fossil fuels unused when limiting global warming to 2°C. *Nature* 517: 187–190.

McKibben, B. 2011. Global warming's terrifying new math. *Rolling Stone*, 2 August.

Nye, D.E. 1996. *American technological sublime.* Cambridge, MA: The MIT Press.

Oil Change International 2016. The sky's the limit: why the Paris climate goals require a managed decline of fossil fuel production. Washington DC: Oil Change International. Available online at http://priceofoil.org/content/uploads/2016/09/OCI_the_skys_limit_2016_FINAL_2.pdf

Rao, S. 2010. Electricity, politics and regional economic imbalance in Madras presidency, 1900–1947. *Economic and Political Weekly* 15: 59–66.

Scholten, D. and R. Bosman. 2016. The geopolitics of renewables: exploring the political implications of renewable energy systems. *Technological Forecasting & Social Change* 103: 273–283.

Shever, E. 2012. *Resources for reform: oil and neoliberalism in Argentina.* Stanford, CA: Stanford University Press.

Smil, V. 2003. *Energy at the crossroads: global perspectives and uncertainties.* Cambridge, MA: MIT Press.

Sovacool, B. 2016. How long will it take? Conceptualizing the temporal dynamics of energy transitions. *Energy Research & Social Science* 13: 202–215.

United Nations. (n.d.) Climate change Available online at http://unfccc.int/paris_agreement/items/9485.php

Valdivia, G. 2007. The 'Amazonian trial of the century': indigenous identities, trans-national networks, and petroleum in Ecuador. *Alternatives* 32: 41–72.

Securities, vulnerabilities and justice

Part 2 explores the consequences and effects of contemporary energy systems. These consequences are socio-economic, environmental and political in nature, and are the out-workings of the energy landscapes examined in Part 1. Securities, vulnerabilities and justice refer to the challenges of energy system transformation. These three terms provide a top-level diagnosis – a meta-language – for the ways in which energy landscapes are currently failing to live up to the demands societies place upon them. Issues of (in)security, vulnerability and justice in relation to the energy system are at the heart of the four imperatives we highlighted in the **Introduction**: to make energy production and consumption more environmentally sustainable; address inequalities in access to modern energy services; ensure energy systems are reliable and secure; and determine how and by whom energy systems should be owned and governed. Accordingly, a set of critical questions run through the chapters in Part 2: for what and by whom is energy used? To whom is energy accessible? For whom, and with what consequences, is energy abundant and made secure? The chapters in Part 2 take up these questions by exploring energy poverty and deprivation (**Chapter 5**); consumption, inefficiency and excess in high-energy societies (**Chapter 6**); the social conflicts and knowledge controversies associated with energy landscapes (**Chapter 7**); and the insecurities and vulnerabilities that surround efforts to reduce the threat of disruption to energy supplies (**Chapter 8**).

A common starting point for the chapters in Part 2 is that demand for energy is about the services (e.g. heating, cooling, power, mobility) that energy provides, and the opportunities for health, well-being, livelihoods and entertainment these energy services make possible. A primary challenge for energy system transformation is that access to these energy services is highly unequal. Billions of people worldwide continue to lack adequate and/or reliable access to modern energy services in both the global North and South. At the same time, a growing proportion of the world's population experiences relative energy abundance, together with the levels of waste and inefficiency that characterise high-energy societies. A related challenge, then, is that the consumption habits of high-energy societies impose significant risks and vulnerabilities, with many (although not all) of these falling on individuals and communities – including future generations – who do not directly benefit from the access to energy services high energy societies enjoy. Climate change exemplifies this relationship at the global scale, but so too do issues like urban air pollution or the disruptive impacts of hydro-electric dams on downstream communities. Moreover, those who lack or have only precarious energy access, and/or who bear the negative consequences of energy systems, frequently have few formal means of redress, and are often on the margins of decision-making processes. 'Justice' increasingly names social demands to address such inequalities in energy access, in the distribution of environmental and other consequences of energy production and consumption, and access to systems of information, participation and decision-making.

A second important idea that runs through Part 2 is a re-worked understanding of the term 'energy system' to reflect our critical social science perspective. In its original engineering and physical science formulation, an energy system refers to the interconnected components involved in supplying energy services. This focus on "all components that relate to the production, conversion, delivery, and use of energy" (IPCC Fifth Assessment Report 2014: 1261) identifies reciprocal relations and interactions among supply, transmission and end use components in the system. Typically, these components and interactions are understood in technical terms: as, for example, the management of load and loss in electricity transmission systems or flow and pressure management in gas pipeline systems. An expanded and 'more social' version of this systems thinking is already in widespread use within energy research, and is described as a 'whole energy system' approach (EPSRC 2017; Parkhill et al. 2013). It maintains the original's core focus on technology, but also incorporates socio-technical, economic and environmental elements in order to address the "the complexities, interactions and interdependencies within the energy landscape and its connections with other systems" (EPSRC 2017). An important point is that the inclusive and integrative ambition of this approach ('whole systems') requires an interdisciplinary perspective able to bring engineers into conversation with economists, sociologists, international relations experts and environmental scientists, for example.

The critical social science perspective of this book builds directly on this whole systems approach. It inherits its symmetrical focus on *both* supply and demand, and its interest in interactions *along* the energy system (i.e. dependencies and feedbacks between energy production and consumption). It also shares its interest in the connections and interactions that take place *between* energy systems and the socio-economic, political and environmental landscape.

However, as a distinctively social science perspective, it pushes a little further to expand and enrich what the 'social' character of energy systems implies. We outlined in the **Introduction** (and elsewhere) some of the ways in which the book seeks to do this, but one of them is particularly significant for the chapters in Part 2. A socio-technical approach understands energy systems as hybrid forms composed of natural resources, material devices, political objectives and cultural norms (amongst other things). The implication of this perspective is that the social character of energy systems is 'hard-wired': it is built-in from the outset, so that it constitutes the energy system in significant ways. The significance of this perspective is that it understands energy systems to be social before they are anything else: issues like technology choice and system design (convention-ally regarded as matters for engineering) are shaped fundamentally by the societies in which these choices and parameters are worked out. For example, what electricity is used for in society and who is (not) served by electricity networks are a product of who participates in decisions about network design and funding, the ways in which societies understand questions of value and worth, and prevailing spatial and temporal frames for decision-making.

The chapters in Part 2 demonstrate that this perspective changes what it means to think about the consequences and impacts of energy systems. Rather than being narrowly the product of (inefficient) technological design, (poor) planning or (inadequate) management, the consequences of energy systems relate to underpinning social structures, institutions, and prevailing cultural practices. From this perspective, for example, energy consumption in high energy societies is not a taken-for-granted outcome of economic and social development, but a product of the structures, norms and practices through which demand for energy is stimulated and reproduced (through, for example, land use zoning regulations that separate work and home, models of economic growth, and cultural expectations around the 'good life'). Within the social sciences, the concept of 'social embeddedness' (drawn from economic sociology and human geography) provides a handy way of expressing this fundamentally social character of energy systems. It articulates a social depth to energy systems largely absent from engineering perspectives, and not sufficiently recognised by the whole systems approach. This book aims to bring it to the fore.

Part 2's focus on securities, vulnerabilities and justice takes forward this distinctive perspective on energy systems. These terms provide an open framing for thinking about the consequences of contemporary energy systems, and the relationship of these consequences to calls for energy system transformation. They offer a top-level set of concepts for thinking about what is at stake in efforts to transform energy systems that illuminates the specific cases and examples highlighted in the chapters. Importantly, each concept also acts as a gateway to significant bodies of research within the social sciences. *Security* draws attention to the social functioning of an energy system and, specifically, and how – and for whom – present conditions and actions reduce future danger and threat (Dalby 2002). In its classic form, it centres on the threat of disruption to energy flows and, in doing so, it also highlights the energy services (and inequalities in energy service provision) made possible by these flows. 'Energy security' has a common-sense appeal to it, yet the term itself says very little about what is done in the name of securing or whose interests are to be served. There is now a large policy-related literature on energy security, as well as a series of more critical social science interventions. These highlight the multiple scales and spaces in which acts of securing energy take place, and the diversity of forms this can take; how efforts to secure energy focus on reducing threats for some, but create insecurities for others (including a range of environmental insecurities, of which climate change is an example); and how turning energy into a security concern often requires mobilising a sense of collective identity (us vs them) in which fear of threat ends up removing decisions from public scrutiny and debate. The plural concept of securities, then, opens the door to a wide range of consequences and gets to the heart of what and for whom energy is for.

Vulnerability describes the "potential for harm and diminished well-being" across a broad range of different dimensions that include health, economic deprivation, life chances, exposures to environmental and technological risks, and marginalisation within/exclusion from energy decision-making (Day and Walker 2013: 28). The term originates in research by human geographers on 'natural hazards' where it emerged as a critical response to explanations of disaster focused only on geophysical causes (flood events, landslides). Vulnerability in this work was a way to explicitly highlight the social processes that made some people and communities more susceptible to harm, and that influenced how hazards affected people in different ways (Wisner et al. 2003). The concept of vulnerability has entered research on the social consequences of energy systems as part of a broad socio-technical perspective. Like security and justice, it can be applied to a range of scales and actors. Here in Part 2 it describes both the deprivations associated with limited access to energy (e.g. energy poverty) and those that arise as a consequence of energy system impacts (e.g. pollution, dispossession, devaluation of property, environmental change). Vulnerabilities is a useful term because it "captures the variability of circumstance and processes" through which these and other harms become expressed, and so conveys the precariousness associated with the potential for harm (Day and Walker 2013: 15). At the same time, it acknowledges important structural axes of social difference that create vulnerability, such as socio-economic position, race, age, disability and gender.

Our use of the term *justice* in Part 2 refers to questions of fairness in society. It draws attention to the outcomes of energy systems and, specifically, to how the benefits and burdens that energy systems create are distributed socially, geographically, and across time. The term is frequently used to draw attention to gross patterns of unfairness: these extend beyond distributive outcomes to also include the procedures governing key areas of decision-making around energy society-relation, notably around recognition (e.g. what forms of knowing – scientific, vernacular, indigenous – are admissible as legitimate knowledge, and whose interests are allowed to count) and participation (e.g. whose voices get to be heard). At the core of the claim for justice is a sense that outcomes and processes of participation can – and should – be made fairer. Calls for justice, for example, are mobilised by social movements contesting inequalities in energy access, proposals for energy infrastructure (mines, dams, pipelines, power stations, disposal sites), and the distribution of environmental and health consequences of energy production and consumption. Part of the value of justice as a concept is that it looks beyond individual preferences to consider the structures and values of the larger social body. A call for justice, then, is an invitation to reflect on the ethical and moral principles that hold society together, and a commitment to the idea that a less unjust society is desirable and achievable (Sovacool and Dworkin 2014). In relation to energy, the language of justice originates in grassroot struggles by activists, non-governmental organisations and concerned communities to document and challenge injustices in, for example, access to modern energy services, the taking of land and water for energy projects, or the consequences of air pollution and climate change. The work of these groups has propelled these energy-related concerns into the academic and policy literature where, in the case of the former, they have combined in interesting ways with previous work on social justice (Pulido 1996; Schlosberg 1999; Timmons Roberts and Parks 2006) and environmental justice (Bullard 1990; Dobson 1998). Justice-talk, then, is now widely used as a way of critically analysing energy–society relations and the goals and promises associated with energy transition (e.g. Bickerstaff et al. 2013).

INTRODUCTION TO THE CHAPTERS IN PART 2

Energy poverty and vulnerability (Chapter 5) explores the relationship between energy and social needs, focusing on patterns, processes and dynamics of energy deprivation. Energy services like heating and cooling can be so fundamental to health, well-being and quality of life that when adequate access cannot be achieved and maintained, serious consequences can result. Chapter 5 outlines the nature and pattern of energy deprivation and highlights the benefits brought about by adequate levels of energy use. It examines the circumstances and processes that produce patterns of energy poverty, along with informal coping strategies and formal policy responses. Overall, the chapter emphasises the systemic conditions that create energy vulnerability worldwide.

Energy consumption, inefficiency and excess (Chapter 6) explores patterns, processes and dynamics of high energy consumption. It illustrates how energy demand is implicated in range of major environmental problems and risks to health, as well as in geopolitical struggles and military conflicts, so that energy demand is a root problem: 'more energy consumption' is not simply a good thing that will enhance living standards. Chapter 6 demonstrates how energy consumption is wrapped up with notions of luxury, comfort, freedom, domesticity and modernity, and how these are mobilised by product producers and retailers, service providers and government actors. A core premise of the chapter is that patterns of consumption are fundamental to system-wide and spatially dispersed vulnerabilities, conflicts and justice issues.

Energy controversies and conflicts (Chapter 7) focuses on the knowledge controversies, social conflicts and political struggles that emerge from, and in turn shape, energy landscapes at a range of scales. It approaches these controversies and conflicts as socio-political consequences of the way energy systems are organised, how they are governed and whom they serve. Conflicts are also symptomatic of how different energy futures are possible, with each allocating the costs and benefits (economic, social, environmental) of energy systems in quite different ways and implying different outcomes for particular communities and social groups. The chapter shows how the uneven social impacts of climate change, chronic urban traffic congestion, oil spills and a range of public health concerns demonstrate many of the contradictions associated with evolving patterns of energy production, distribution and use. A key focus of the chapter is on the underlying processes that generate forms of social conflict around energy questions, and the politics surrounding scientific knowledge claims, ethical and moral claims, and claims for rights of participation and recognition.

Energy securities (Chapter 8) highlights how energy security has become a key policy framing, and how its mainstream deployment privileges some material sites and horizons of action over others: for example, energy security is frequently framed as a question of resource supply and a matter of national sovereignty. The chapter explores the limits of this framing and how the nexus of energy and security occurs in settings and at scales other than the nation state (e.g. region, city, household). By asking 'secure for whom?' the chapter examines the social and political consequences of making energy secure across a wide range of settings. Attention to alternative scales and practices (e.g. city-based initiatives to 'secure' critical inputs to urban living, off-grid communities) shows how people and places can become vulnerable and energy insecure, not least because of the embeddedness of energy in different economic and social practices. Overall, the chapter considers how the concept of energy security is implicated in the reproduction of existing power relations at a range of scales.

REFERENCES

Bullard, R.D. 1990. *Dumping in Dixie: race, class, and environmental quality*. Boulder, CO: Westview.

Dalby, S. 2002. *Environmental security*. Minneapolis, MN: University of Minnesota Press.

Day, R. and G. Walker. 2013. Household energy vulnerability as 'assemblage'. *Energy justice in a changing climate: social equity and low-carbon energy*. K. Bickerstaff, G. Walker and H. Bulkeley (eds), London: Zed Books, pp. 14–29.

Dobson, Andrew. 1998. *Justice and the environment: conceptions of environmental sustainability and dimensions of social justice*. Oxford: Oxford University Press.

EPSRC. 2017. Whole energy systems. Engineering and Physical Science Research Council, London. Available online at www.epsrc.ac.uk/research/ourportfolio/researchareas/wholesystems/

IPCC 2014. Climate Change 2014 Mitigation of climate change: Working Group III Contribution to the Fifth Assessment Report of the Intergovernmental Panel on Climate Change. Cambridge University Press. Avilable online at www.ipcc.ch/pdf/assessment-report/ar5/wg3/ipcc_wg3_ar5_full.pdf

Parkhill, K., C. Demski, C. Butler, A. Spence and N. Pidgeon. 2013. *Transforming the UK energy system: public values, attitudes and acceptability – synthesis report.* London: UKERC.

Pulido, L. 1996. *Environmentalism and social justice: two Chicano struggles in the southwest.* Tucson, AZ: University of Arizona Press.

Schlosberg, D. 1999. *Environmental justice and the new pluralism: the challenge of difference for environmentalism.* Oxford: Oxford University Press.

Sovacool, B.K. and Dworkin, M.H., 2014. *Global energy justice.* Cambridge: Cambridge University Press.

Timmons Roberts, J. and B. Parks. 2006. A climate of injustice: global inequality, North–South politics, and climate policy. Cambridge, MA: MIT Press.

Wisner, B., P. Blaikie, T. Cannon and I. Davis. 2003. *At risk: natural hazards, people's vulnerability and disasters.* Abingdon and New York: Routledge. Second edition.

Energy poverty and vulnerability

5

Energy can be so fundamental to human well-being that serious consequences may result when adequate access to its associated services cannot be achieved. This chapter examines the circumstances and processes through which patterns of energy deprivation are produced in different parts of the world, focusing in particular on the settings of home, household and community. The main framings through which problems of inadequate energy access have been examined to date are introduced (energy and fuel poverty, vulnerability) while drawing out their differences and connections; the geographical variability in how these problems are experienced and understood is also stressed. The chapter also examines the different consequences of inadequate access to energy services (and their geographies), in addition to providing a critical interrogation of the informal and formal amelioration strategies that are utilised and developed in this context. This leads to a discussion of the relationship between energy equity and key determinants of quality of life, so as to highlight the benefits brought about by adequate levels of energy use. Our review of energy poverty and vulnerability is situated within a broader consideration of the political, institutional and infrastructural underpinnings of marginality in the global North and South alike. We recognise the problems involved in using dualisms like developed/developing, rich/poor and even global North/global South. However, alternative formulations (like low-and-middle income countries) also have significant limitations. For this reason, we have adopted global North

energy access

seen as a key premise for alleviating energy poverty, understandings of energy access have primarily depended on the availability of organised infrastructural systems for providing different types of energy. Energy access often hinges on electrification, itself associated with the introduction of electric power to a given location. Historically, the development of electricity grids has been seen as a key vehicle for energy access: more recent thinking around energy access extends to the provision of decentralised off-grid and electricity storage solutions.

energy affordability

one of the main contributing factors to fuel and energy poverty. It expresses the ratio between household incomes and energy prices: a rise in energy prices in relation to household income makes the relevant energy carrier less affordable. Energy affordability is sometimes expressed via the 'energy burden', which equals the percentage share of household energy expenditure within its total income (this is sometimes calculated in 'equivalent' terms, when both the expenditure and income figure are divided by the number of household members weighted by age). A higher energy burden is a sign of fuel or energy poverty.

fuel and energy poverty

in the UK, fuel poor households were initially defined by the government as those needing to spend more than 10% of their total household income before housing costs on all fuel used

and South in this chapter and throughout the book as a shorthand for describing enduring structural inequalities at the global scale (for a reflection on the politics of labels, see Silver 2015).

In its entirety, the chapter aims to connect the causes and consequences of energy deprivation with current developments at the global scale. Using energy services in the home as a lynchpin, it outlines the key components that contribute to inequalities in energy consumption: questions of **energy access**, **energy affordability**, energy efficiency and household needs. The chapter also discusses the dominant definitions of **fuel and energy poverty**, while outlining their comparative advantages and shortcomings. Sensitive to the profoundly different circumstances of energy deprivation in different parts of the world, the chapter aims to move beyond the overly simplistic dualistic global North–global South framings that dominate mainstream understandings of the underlying driving forces and socio-economic implications of the dynamics of energy inequality. In doing this, the chapter builds on the notion of multiple/overlapping rhythms and technologies of energy use in the home, initially suggested in **Chapter 3**.

DIMENSIONS AND FRAMINGS OF ENERGY POVERTY

In countries of the global North, problems of energy deprivation in the home are most commonly described via 'fuel poverty' frameworks. The basic features, driving forces and consequences of fuel poverty were established by Brenda Boardman (1991), in her seminal book titled *Fuel poverty: from cold homes to affordable warmth*. Boardman's approach is predicated upon the notion that fuel poverty is additional to general income poverty, as it involves the energy efficiency of the built fabric of the home, including walls, roofs, windows, appliances and heating systems. This is because fuel poor households are purchasing warmth that is 'unaffordable' due to the loss of 'useful' energy during conversions from fuel inputs to final energy consumption that takes place in the home. Boardman's work, like that of many of her successors in the field, was focused on the UK and Ireland: countries that are characterised by high rates of fuel poverty due to the combination of high rates of income inequality and inefficient housing. In such states, fuel poor households tend to be concentrated in homes that are 'hard to heat' due to insufficient levels of insulation: a historical legacy dating back to the mid-nineteenth century, when buildings were built to low-quality standards, and there was an emphasis on ventilation rather than thermal comfort. It is worth noting that cavity wall insulation did not become common in the UK until well into the twentieth century, with double glazing in domestic buildings becoming an explicit standard only in 1994, even though it had been a recognised measure since 1857.

In essence, fuel poverty scholarship has helped challenge the suggestion that this predicament is reducible to income, price or other economic factors, as exemplified by the statement of a UK government minister in the 1980s that "People do not talk of 'clothes poverty' or 'food poverty' and I do not think it is useful to talk of 'fuel poverty' either" (Campbell 1993: 58). The existence of a close relationship between energy efficiency and fuel poverty foregrounded the importance of housing stock characteristics, because the level of thermal insulation and the nature of heating systems are highly dependent on the age, type and quality of the built fabric of the home. This brings into focus the highly geographic nature of domestic energy deprivation: even though fuel poverty is primarily felt within the confines of the home, the fact that its depth and extent are influenced by wider residential and infrastructural formations means that the condition exhibits significant geographical variation across a much wider range of material sites. Neighbourhood, city regional and country scale variations in the quality of housing are reflected in the emergence of an uneven geography of fuel poverty, in which residential dwellings combine with broader patterns of social inequality and energy prices to create a distinct spatial distribution of domestic energy deprivation. Vulnerabilities to fuel poverty may be concentrated in particular areas with low-income households or poor quality housing, or diffused within relatively better-off urban or rural districts.

More recent debates on the driving forces of fuel poverty have focused on a wider range of dimensions beyond incomes and housing. There has been an increased focus on domestic energy needs, which lie at the heart of household **energy vulnerability**: residential energy deprivation occurs when the heating, lighting and other energy requirements of the occupants of a given home are unmet (Table 5.1) (Bouzarovski and Petrova 2015). As a result, it has become apparent that households who demand above average amounts of energy may find themselves more vulnerable to domestic energy deprivation; common examples include pensioners and families with small children. Moreover, energy needs are largely socially and spatially constructed, as they are tied into wider expectations about what constitutes a well-functioning home and household. This means that vulnerability may arise as a result of excessive pressure to maintain standards of consumption that are economically and technically unattainable within a given set of individual circumstances. A mismatch between energy needs and housing stock formations may result in the emergence of fuel poverty even among financially better-off households living in well-insulated dwellings. This is particularly true in cases where houses are infrastructurally hard-wired into particular types of heating systems that increase energy costs while failing to meet the specific needs of the residents concerned (see **Example 1**).

to heat their homes to an acceptable level. A more recent definition considers households fuel poor if they have above-average fuel costs which, if spent fully, would leave them with a residual income below the official poverty line. Wider understandings of energy poverty see this condition as one in which households cannot secure a socially and materially necessitated level of energy services in the home.

energy vulnerability encompasses the risk factors that underpin an entity's propensity to experience a lack of needed energy supply. This can refer to both nation states, regions and cities; as well as households, where the lack of adequate energy services is commonly tied to the emergence of fuel and energy poverty as well as blackouts or brownouts.

Table 5.1 Deprivation issues associated with different domestic energy services

Type of energy service	Relevance for domestic energy deprivation
Space heating	Principally a problem for households in cold climates – in countries of the global North this includes low income groups, as well as those living in inefficient homes. Gaining access to more effective, comfortable or efficient methods of domestic heating is associated with the situation in countries of the global South. The two problems may combine in countries with relatively mild winters where households do not have adequate heating systems in their homes – in Europe, the highest numbers of excess winter deaths are found in countries like Cyprus and Portugal.
Water heating	Poverty implications mainly discussed in relation to global South, although this service is a significant component of energy consumption in countries in the global North.
Space cooling	An energy-related problem for households living in climates with unbearably hot summers, and urban areas in particular. Climate change-related heatwaves have exacerbated the issue. Both the inadequacy and access to the service may be an issue. Research on global South contexts is lacking beyond large-scale models.
Lighting	A global domestic energy deprivation-related challenge. Mainly researched in the context of the global South where it is associated with a lack of access to electricity. A reduction of indoor lit spaces in relation to affordability issues has also been observed.
Cooking	Most of the literature explores this service in relation to energy poverty in the global South, where the lack of access to electricity is a major obstacle towards economic development and well-being.
Drying	Rarely connected to energy deprivation, although the lack of adequate facilities for this service in colder climates (whether provided by networked infrastructures or not) has been connected with adverse health impacts.
Refrigeration/ appliances/IT	These services are directly linked to the affordability and availability of electricity infrastructures, and as such can be found across the world. Levels of consumption are culturally and socially conditioned, which means that deprivation 'thresholds' are highly context-specific.

Source: Based on Bouzarovski and Petrova (2015)

The introduction of concerns around energy needs, infrastructures and vulnerabilities has allowed discussions of developed-world energy deprivation to extend beyond their traditional home in the UK and Ireland onto a wider range of geographical contexts. The term 'energy poverty' has replaced 'fuel poverty' in such settings, signalling a conceptual move away from the supply of fuels to the broader systems and networks of energy provision in the home. Of particular note is the situation in the post-communist countries of Eastern Europe, where domestic energy deprivation is widespread due to the combination of fuel price increases, inefficient housing, and inadequate state policy. In the US, energy poverty has been contingent on the legal and regulatory arrangements that govern the country's fragmented utility market, with households living in areas of high prices being particularly vulnerable. Many American homes were built in periods of fossil fuel abundance, resulting in material and social arrangements that disfavour energy conservation and efficiency in the home (Harrison 2013). Yet, energy poverty is also present in countries such as Germany and Austria, whose housing stock, on average, is

---

EXAMPLE 1: ENERGY VULNERABILITY AND INFRASTRUCTURAL LOCK-IN

Households can find themselves 'trapped' in situations of domestic energy deprivation due to their housing arrangements and/or everyday needs. Technical constraints or physical obstacles in the fabric of buildings can prevent households from switching to more affordable or efficient sources of energy; the occupants of homes in apartment buildings that are equipped only with night storage heaters, for example, often find themselves facing above-average energy costs and, at the same time, are unable to switch to cheaper heating alternatives. An additional problem is that night storage heaters are not well suited for households with energy needs that reach their peak in the evenings, as the heaters' ability to release useful energy decreases during the day. In a similar way, district heating systems – if configured in ways that provide only limited opportunity for user-control – may lead to the exacerbation of energy poverty (Tirado Herrero and Ürge-Vorsatz 2012).

---

characterised by some of the highest energy efficiency levels in the world. In such states, the condition tends to affect low-income households with arrears and debts in the payment of energy bills.

The concept of 'energy poverty' has also been used to describe the circumstances faced by developing-world countries, whose inhabitants often face a sheer absence of infrastructures for the delivery of **modern energy services** to the home. Termed the 'other energy crisis' by some, this issue has been high on the agenda of international development organisations and donor agencies, while receiving significant political and economic attention by a growing number of governments in less-developed economies. It is evidenced by the UN's Sustainable Energy for All initiative, which gives prominence to the impact of international and local efforts on the provision of energy services at the local scale, while foregrounding the need for integrated thinking to address concerns of climate change, natural resource scarcity, and global income inequality (Mahama 2012). The initiative was developed partly in response to the omission of energy from the UN Millennium Development Goals (MDGs), which were established in 2000. This absence has subsequently been addressed by the inclusion of a specific goal targeting affordable and clean energy (Goal Number 7) within the 17 Sustainable Development Goals (SDGs) agreed by the international community in 2015.

These global energy poverty alleviation efforts aim to address the circumstances faced by the estimated 1.2 billion people who still do not have access to electricity in their homes, and the 2.8 billion who are forced to rely on traditional biomass for cooking and heating purposes. A map of energy poverty-related issues at the global scale shows a degree of differentiation between Africa and Asia with regard

modern energy services
the UN's Sustainable Energy For All (SEforALL) initiative, and the energy-specific Sustainable Development Goal Number 7, aim to deliver modern energy services to all global citizens by 2030. There is some debate over how these modern energy services should be defined. The definition employed by SEforALL's global tracking framework defines modern energy services as (i): electricity access (availability of an electricity connection at home or the use of electricity as the primary source for lighting); and (ii) access to modern cooking solutions (defined as relying primarily on non-solid fuels for cooking). These definitions leave unanswered significant questions about levels of access, reliability, and the costs involved in securing sufficient access to these modern energy services.

to access to electricity and access to clean cooking facilities (Figure 5.1). Given the deficiency of basic systems of modern provision that underpin such circumstances (Figure 5.2), energy poverty policies have traditionally been focused on expanding and improving the supply of fuels and the construction of new large-scale infrastructures. In particular, there has been an emphasis on the electrification of areas beyond the reach of existing power grids, whether via expansion of national grid systems or a range of potential off-grid solutions (see below). Experience from those countries across the global South that have successfully enhanced electricity access amongst their populations suggests that electrification can help move the inhabitants of rural areas out of poverty, by dismantling some of the principal obstacles to improving livelihoods while improving access to education and the overall quality of life (although there are also many counter examples that stress that the relationship between socio-economic outcomes and physical access to electricity is far more complex than this might suggest). International financial institutions such as the World Bank, Asian Development Bank, African Development Bank and the Inter-American Development Bank have continued to spearhead electrification initiatives under the SEforALL initiative, often as part of wider programmes for supporting economic growth and frequently via the promotion of private sector-led solutions within the energy sector.

However, the use of conventional electricity supply-orientated logics in the global South – relying on centralised generation and power line distribution networks – has been criticised for favouring the interests of large corporations and political elites while failing to address the everyday needs of vulnerable people. For example, it has been argued that the expansion of generation

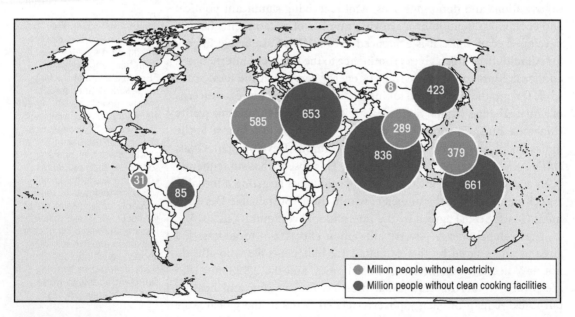

Figure 5.1 Inequalities in access to energy at the global scale (based on IEA 2011)

Figure 5.2 Beneath a brightly lit Delhi, India, metro station, a community of 200 people live in extreme energy poverty, with one illegal line bringing electricity into a single tent (photo credit: Marilyn J Smith)

technologies (such as big dams) aimed at increasing electrification rates in developing countries has often resulted in deleterious social and environmental impacts for local communities (see Lee et al. (2016) and **Example 2**). Even where communities are connected to the national grid, paying for exorbitant connection charges can remain a significant challenge can for the poorest sectors of society. The poorest communities are also often forced to put up with the worst service and the most disruptions to supply. Gender aspects have also received inadequate attention in the classic energy development paradigm, despite widespread evidence that the economic, labour and health burdens of energy poverty are disproportionately borne by women of all generations (Pachauri and Rao 2013).

It is for such reasons that scientific and policy attention has increasingly turned towards the poverty-reducing potential of microgeneration and renewable energy investment, often as part of a wider process of **energy leapfrogging**. Practical efforts to provide local communities with investment tools and development know-how for alternatives to top-down power grid expansion have been accompanied by a rising awareness of how household energy transitions towards modern fuels in the developing world are strongly determined by cultural and political factors. These factors affect the sustainability of off-grid initiatives as much as they do grid extension programmes, as studies of the gendered

energy leapfrogging
an optimistic hypothesis that countries in the global South may move more easily onto low-carbon forms of energy and recovery as a result of being able to access energy-efficient technologies that were not available during past energy transitions. The expectation is that transfer of technology and know-how from the global North will allow such states to take up a trajectory of low-carbon development without first increasing the energy intensity of the economy. A similar argument has been used in the case of former communist states in Europe and Asia, which inherited high levels of public transport use, network heat provision and urban density from communism; and as such could have directly adopted more sustainable forms of energy in the process of implementing a market capitalist economic model.

EXAMPLE 2: LIVING UNDER THE GRID IN THE GLOBAL SOUTH

The international community has decisively woken up to the importance of tackling energy poverty over recent years. Evidence for this includes global commitments to electrification, decarbonisation and the spread of clean cooking technologies, as part of the UN Sustainable Energy for All (SEforALL) Initiative and the Sustainable Development Goals. Many governments remain wedded to the idea of national grid extension as the most effective strategy for achieving energy access objectives. Research is beginning to reveal the limitations of such strategies, however, by highlighting how access can remain extremely limited even where the grid is extended to rural communities. For example, Lee et al. (2016) show in their study of electrification in Kenya how "even in a seemingly ideal setting, where there is high population density and extensive grid coverage, electrification rates remain very low, averaging 5% for rural households and 22% for rural businesses". Highlighting the extent of this exclusion, the researchers note that half of the 'unconnected' households in their study were located directly beneath transmission wires (i.e. were 'under-grid'), or clustered within 200 m of a low-voltage power line. A similar pattern can be see with the development of national gas transmission pipelines and 'gas grids': a common complaint of communities in eastern Russia, for example, has been that natural gas extracted in the region, or transiting through it via long-distance pipelines, effectively bypasses local demand for modern energy services in order to service demand elsewhere.

---

impact of some off-grid initiatives reveal (Standal 2016). It is becoming increasingly accepted that questions of access to modern energy cannot be addressed without considering the broader institutional and power relations that underpin the circulation of energy in society and its absence, as well as associated dimensions of distributional and procedural justice. Understandings of energy poverty, therefore, have shifted from supply-dominated approaches focusing on the expansion of technical infrastructures only onto the systemic forces that have allowed the condition to emerge and persist (Sovacool 2012).

A possible entry point into the integration of developed- and developing-world perspectives on domestic energy deprivation is the suggestion that deprivation universally (i.e. regardless of location or context) undermines security of access to 'energy services' in the home. Energy services are the "benefits that energy carriers produce for human well being" (Modi et al. 2005: 9). The term acknowledges how people do not demand energy in and of itself, but seek the benefits that energy can bring to everyday life, including mobility, heating, cooking, cooling, lighting, information, entertainment and power for the utilisation of machinery (see **Chapter 1**). An emphasis on energy services in the home shifts the focus of policy debates away from the quantitative delivery of energy carriers – coal, oil, electricity or gas – onto the qualitative achievement of adequate domestic functions. However, energy services themselves hinge upon subjective factors (Petrova et al. 2013) as they are orientated towards the meeting of household and community energy needs which, as pointed out above, are themselves culturally and socially determined. Moreover, the systems through which energy services are provided to households involve "different inputs of energy, technology, human and physical capital, and environment (including natural resources)" (Haas et al. 2008: 4013). In short, energy services involve complex interactions between technologies and final users: they can be thought of as hybrid 'assemblages,' made up of different technical, material and social components, and operating across multiple scales and sites that extend beyond the confines of the home (Bouzarovski and Petrova 2015). Thus, policies aimed at reducing or eradicating energy deprivation need to consider the wider series of political, economic and cultural factors influencing the chain of transformations that lead to the production of useful energy in the home and the uses to which energy is put (and hence the broader socio-economic outcomes that addressing energy deprivation can deliver within different circumstances).

The idea of energy services lies at the core of the **energy ladder and stacking** model. This widely used model suggests the technological sophistication of fuels used in the home increases hand in hand with household income (Figure 5.3). The implication of such approaches is that energy poverty reduction policies should aim to achieve a one-directional transition towards particular types of energy use. But the concept of a simple 'energy ladder' has been criticised for failing to

energy ladder and stacking
a theoretical model that posits that households – and the wider urban, regional and national economies that they are part of – will use progressively more technologically advanced energy carriers as their incomes increase (Figure 5.3). A corollary of such thinking is that 'household fuel transitions' are associated with economic development. At the same time, it is argued that households with higher incomes also tend to 'stack' fuels by relying on a greater variety of energy carriers.

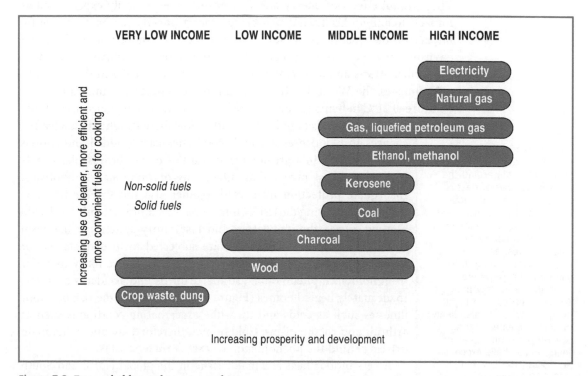

Figure 5.3 Energy ladder and energy stacking

capture the diversity of house-hold circumstances: these circumstances are important because they often mean different fuels are used simultaneously in the home due to social preferences, affordability or efficiency levels. For example, the use of fuel wood and wood-burning stoves may be characteristic of 'low income' households, but households with higher incomes may still choose to retain this fuel and energy conversion technology: a wood-burning stove can provide a feeling of cosiness and comfort, and offer a visual focus, in addition to more conventional functions such as space heating, hot water, cooking, drying and light. Furthermore, richer families in urban settings may continue to utilise traditional stoves alongside more modern fuels because of their suitability for preparing different types of food and issues of preferred taste: Mekonnen and Köhlin 2008). At the same time, energy carriers associated with different parts of the energy ladder model often provide similar energy services in a single home: for example, illumination can come from candles, oil lamps, animal dung or electricity. The energy ladder has thus been superseded by a 'fuel stacking' model, which aims to capture the existence of a combination of fuels and technologies at different income levels.

WELL-BEING, HEALTH AND ECONOMIC DEVELOPMENT: ADDRESSING DOMESTIC ENERGY DEPRIVATION AND ITS CONSEQUENCES

The harmful effects of energy and fuel poverty are primarily experienced as impacts to human health and well-being. Energy poverty can have a powerful impact on the quality of domestic life (Figure 5.4). Indoor air is the main environmental factor linking energy deprivation to poor health, mainly because the former affects ambient temperature levels and air quality in the home. In cold climates, the World Health Organization recommends a minimum temperature of 21°C in living rooms, and 18°C in all other residential spaces. Acute exposure to levels of indoor warmth below these thresholds can lead to changes in blood pressure and blood chemistry (Liddell and Morris 2010), resulting in a greater risk of cardio- or cerebrovascular events such as strokes, myocardial infarctions or pulmonary embolisms. Inadequate protection from cold weather also impairs the immune system, intensifying the risk of infections. At a more systemic level, cold housing affects the circulatory and respiratory systems: it has been established that households that are subjected to inadequate indoor temperatures for prolonged periods of time are more likely to suffer from influenza, pneumonia, asthma, arthritis and accidents at home. Inadequately heated homes (Figure 5.5) also increase the risk of minor illnesses such as colds and flu while exacerbating conditions such as arthritis and rheumatisms. Cold homes, therefore, are one of the main drivers behind the phenomenon of **excess winter deaths**.

Excess indoor heat is a major issue in the global North and South alike. Domestic energy deprivation is known to cause significant health

excess winter deaths
the difference between the number of deaths that have occurred in winter and the average number of deaths during the preceding four months and the subsequent four months. This number is important in countries with cold climates, as approximately one-fifth to one-ourth of EWDs are attributed to energy poverty. Thus, countries with higher rates of EWDs are also likely to have a greater energy poverty problem. In recent years, EWDs have exceeded 30,000 in the UK and 100,000 in the US.

Figure 5.4 Indoor air pollution from traditional cooking methods (photo credit: Practical Action/Crispin Hughes)

Figure 5.5 Coal stove in a house in Budapest, Hungary, the residents' sole heating source (photo credit: Neil Simcock)

impacts in cases where indoor temperatures exceed particular limits during hot summer weather and, more rarely, as a result of prolonged exposure. In the United States, excess indoor heat has been the leading weather-related cause of death for several decades. The issue has a clear geographical dimension, as it has been shown that heat-related health effects are linked with income differentials within cities and neighbourhoods (Uejio et al. 2011). While global research on this subject is limited, spikes in the number of deaths during heatwaves in countries such as France, Australia and the UK testify to the need for adequate space cooling in addition to space heating – a requirement that is likely to increase as average temperatures and weather extremes increase as a result of climate change. There is evidence that fuel and energy poverty also affect mental health problems such as anxiety and depression, particularly among adults and adolescents. Educational attainment among children is also adversely impacted, due to school absences or the lack of adequate study places in the home. The effects of inadequate access to energy services in the home are particularly pronounced in developing-world countries. Here, the effects of deprivation extend beyond the limits of the home to encompass the wider landscape of amenity provision in communities, involving a range of infra-structural, social and economic services. In this way energy deprivation can lead to a wholesale restriction of opportunities for education, enterprise and investment.

The link between energy poverty and health also functions via indoor air pollution. This problem is particularly pronounced in developing-world countries, where households are likely to use open fires or stoves without operating chimneys and hoods. There are estimates that more than 1.5 million people per year suffer from premature deaths as a result of inadequate biomass combustion-related air pollution. This is mainly due to the excessive emission of particulate matter, as well as substances such as carbon monoxide, sulphur oxides, nitrogen oxides, aldehydes, benzene, and polyaromatic compounds. At the same time, the collection of solid fuels itself carries a number of health and safety risks, in addition to time opportunity costs. Dependency on local biomass can lead to land degradation by fuelling deforestation, soil erosion, biodiversity loss and local water pollution. The process may change existing patterns of vector-borne disease transmission, leading to further impacts on health. Where solid fuels are not available, however, households may resort to the use of plastic or other waste, the combustion of which releases an entirely different array of toxic pollutants. Interestingly, even in the case of the accelerated take-up of the improved cookstoves so heavily promoted by international agencies and NGOs over recent years, research has suggested that their impacts on reducing health impacts may be far lower than previously anticipated (see Mortimer et al. (2016) on this, as well as Batchelor (2015) on the potential for electric cooking alternatives).

The central importance of energy services for the conduct of everyday life means that the health and well-being impacts of energy deprivation extend beyond ambient temperatures and air pollution. Energy deprivation can limit access to resources and amenities that are socially necessary, such as domestic appliances (including television and radio sets) and information communication technologies (ICTs). In addition to requiring access to reliable electricity supplies, the operation of such devices and systems hinges upon a host of infrastructural preconditions that demand extensive technological investment and organisational efforts. The absence of these conditions in particular geographical settings has fuelled the emergence of a 'digital divide' – commonly described as the gap between people and places who are able to utilise the benefits of ICT, versus those for whom access is restricted or impossible. The digital divide has been shown to exist between the inhabitants of cities and rural areas, while also being defined by socio-economic status, education levels and gender. At the global scale, many countries in the global South find themselves on the unfavourable side of this split. The divide may be present even in cases where infrastructural access to ICT has already been established, due to the inability to afford higher-performance computer hardware, a lack of high-speed internet connections, and the limited availability of subscription-based content. In some cases, the absence of ICT and appliance services can contribute to the rise of energy or fuel poverty itself. For example, in many countries with liberalised retail energy markets – i.e. where customers choose between different suppliers – the lowest-cost gas and electricity deals can only be obtained by switching between different gas or electricity suppliers on the internet. This may be impossible for low-income income or other vulnerable groups who do not utilise or cannot access the necessary levels of ICT. Addressing the digital divide in energy services, therefore, can yield direct economic benefits, in addition to broader evidence that it can also help to improve literacy, democratic participation and social mobility.

Fuel and energy poverty affect individuals and communities worldwide and, as a consequence, there are multiple approaches for addressing or managing the condition. A useful distinction for thinking about these different strategies for dealing with energy deprivation is between, on the one hand, a physical lack of availability of an energy carrier like electricity (which may be a permanent condition or a function of unreliable supply); and, on the other, an economic constraint on accessing physical supply due to its affordability. Households that do not have access to electricity address this predicament by using alternative fuels or infrastructural amenities, in spite of the health impacts described above. Provided they can afford them, such families may buy petrol- or diesel-fuelled generators for domestic electricity production; those with more limited budgets tend to use rechargeable solar lamps. The same technologies are used in situations where the electricity supply is intermittent or unreliable, often in conjunction with batteries that can be charged during periods of power connectivity (see **Case study**). At the same time, the lighting needs of low-income households in the global South are overwhelmingly met by candles and kerosene-powered paraffin lamps. Bottled gas (LPG), charcoal and biomass are the main energy carriers used for cooking in such households. An income gradient exists among these fuels, with biomass and charcoal more common

CASE STUDY: LIVING OFF-GRID – SOLAR HOME SYSTEMS IN BANGLADESH

Contributed by Raihana Ferdous, Durham University, UK

Bangladesh has one of the lowest electricity consumption rates in the South Asia. Consumption per person (310 kWh) has nearly doubled in a decade, but is still less than half that in India and only one-tenth of the world average. Around 60% of people in Bangladesh have access to electricity, although the proportion is significantly lower in rural areas. Homes, offices and factories that do have a connection to the national grid frequently suffer interrupted and unstable power supply, due to the combination of out-dated equipment, poor maintenance and corruption in the energy sector. Power outages and load-shedding are a feature of urban life.

In this context, off-grid 'solar homes' fitted with photovoltaic cells have been extensively promoted as a clean, affordable and decentralised way of providing access to electricity in both rural and urban areas. Bangladesh now has one of the fastest growing solar home systems in the world (Urmee et al. 2016). Over 4 million solar home systems have been installed in off-grid areas where electrification through grid expansion is challenging and costly. The programme aims to support 6 million solar homes in the next few years, with a combined generating capacity of 220 MW. The off-grid solar home programme aligns with the national government's policy of providing electricity for all by 2021 – the fiftieth anniversary of Bangladesh's independence – with a goal of 10% of electricity coming from renewable energy sources. It is also shaped by the global Sustainable Development Goals (specifically SDG7 on energy access) with the programme delivered through public–private partnerships and funded with around US\$ 700 million in loans and grants from international donor organisations. At the same time, the government is pursuing other, more conventional responses to the lack of electricity access and poor-quality infrastructure. The state-run Bangladesh Power Development Board is undertaking a major expansion of generating capacity (8,000 MW), which includes incentivising new power station construction (coal, gas and nuclear) and importing electricity from neighbouring countries.

The solar home system is considered a great success, in terms of the number and pace of installations. There are, however, several limitations to the system: simultaneously social and technical in nature, these affect for what, by whom and where solar electricity is used. First, the solar home system energy is not able to meet one of the most important energy demands, which is for cooking. The inability of solar to replace traditional cooking fuels like wood and dung means poor people continue to be exposed to serious health and environmental consequences. Second, power and storage issues constrain how and for what solar electricity can be used. A typical system comes with four hours of battery backup, sufficient to power a few light bulbs and a mobile phone charger. Some larger (and more expensive) systems are able to power small home appliances such as televisions. The problem of power storage means the system has only limited capacity to support income-generating activities in off-grid areas. For example, the system does not provide an adequate electricity supply for performing medical tests and operations, or even to run a refrigerator for storing medicines: despite having solar energy at home or in their immediate community people living in off-grid areas still need to travel to access medical services. Third, although there are multiple sources of financial support for the solar home system, the price of solar electricity is still four to six times higher than grid-connected electricity (the cost of which is subsidised). As a consequence, a large part of population is unable to afford solar energy and so the solar home system has had only a limited impact on poverty alleviation (Bhattacharyya 2006).

Overall, the solar home system in off-grid areas provides mostly recreational and leisure benefits as well as a certain level of social status (Rahman and Ahmad 2013). The programme has successfully created a consumer group in off-grid areas: it has eliminated darkness in the private space of the home, and connected off-grid communities able to afford it with the rest of the world, via mobile phone and digital media. But the solar home system has done little for public services like street lighting, public safety and security. Roads and streets are still dark in the off-grid areas, which restricts easy movement especially for women and children. The solar home system's socio-technical and spatial limits are reflected in the way that people in off-grid areas distinguish between 'solar' and 'line' (i.e. grid electricity). Many of those who use solar home systems consider solar energy to be inferior to grid electricity and they continue to want 'line': for them solar is not an alternative to grid connection.

A short documentary film based on this research, illustrating off-grid life in rural Bangladesh, can be found online at https://vimeo.com/121374407.

among less well-off groups: bottled gas, where available, is typically only used by higher-income households. A number of countries have undertaken large-scale programmes to provide rural households with LPG for cooking, in order to replace more polluting and energy intensive fuels such as charcoal and biomass. Such measures offer alternatives in situations where the development of large-scale networked infrastructures for gas and electricity is financially, organisationally or technologically challenging. The provision of small-scale renewable energy also offers a way out of energy poverty, while opening opportunities for 'leapfrogging' the traditional pathways of infrastructural development followed by developed-world countries. However, in comparison to the frequently subsidised costs of grid-provided electricity, the costs of such schemes may result in the rural poor paying more for electricity than grid-served communities. There also remain question marks over the potential of off-grid energy to provide real opportunities for livelihood enhancement (see **productive use**).

A different set of strategies are at play in instances where the affordability, rather than the availability, of adequate energy services is a source of hardship for households and communities. Such situations are characteristic of countries of the global North with a need for space heating in winter, although they are also faced by urban populations across the global South. At the same time, it is

productive use

household solar home systems have spread rapidly as a way of meeting the energy access needs of poor off-grid communities across the global South. A criticism, however, is that these systems generally only provide access to basic lighting, mobile charging and perhaps some household entertainment (radio/ television), in contrast to the potential use of grid or community-scaled mini-grid systems for more intensive productive uses, such as refrigeration. The more restricted productive uses to which electricity from solar home systems can be put has led some to argue their role in addressing energy poverty is limited. Currently, considerable effort is being expended into making community energy systems more economically viable, and expanding the size of individual household systems and the range of appliances that they can power.

increasingly recognised – in part as a result of the weather extremes and intensifying summer heatwaves brought about by climate change – that space cooling may also be an important issue in this regard. Reducing room temperatures, coupled with wearing additional clothing, are the principal practices employed by households for 'coping with the cold'. The opposite set of measures are implemented in combating excessive heat. Energy conservation measures that require little or no capital investment are also commonly used – this can involve sealing skirting boards, using draft excluders, drawing curtains, or closing window covers. As far as other energy services are concerned, electricity can also be saved by turning appliances off rather than leaving them on standby, unplugging chargers, switching lights off when not in use, not overfilling kettles, putting lids on saucepans when cooking, reducing temperature settings and running full loads on washing machines, driers and dishwashers, as well as using lines to dry washing. Households also address fuel poverty by utilising social networks and community resources, both in terms of the informal exchange of energy-related goods and services (e.g. lending money or sharing an electricity lead) and the use of public heated or cooled spaces (e.g. libraries, pubs or even friends' houses). During periods of extreme cold or heat, state authorities may provide public spaces for vulnerable individuals or households to seek shelter.

Energy conservation and sharing measures provide immediate relief to groups affected by domestic energy deprivation. A long-term resolution of such challenges, however, cannot be achieved without capital investment in the energy efficiency of the built fabric, appliances or heating systems. This can range from relatively low-cost steps such as switching to low energy light bulbs and appliances, to more ambitious measures involving the installation of more efficient boilers and the insulation of lofts, windows and cavity walls. The high upfront cost of such activities means that low-income households are usually unable to afford them – a situation further exacerbated by the poor quality housing that such groups inhabit, and their high utility costs. In a number of countries, governments or utility companies have been supporting energy efficiency investment among vulnerable groups via targeted programmes funded by general taxation or levies on energy bills.

CONCLUSION

The common theme uniting widely different contexts and geographies of energy and fuel poverty – whether urban or rural, or in the global North or global South – is that each situation is the result of unfulfilled energy needs. The desire of human beings for energy services is, in part, biologically determined: acute or chronic health impacts start to occur below and above certain thresholds of temperature or air quality, so that the provision of adequate energy services, all too often, is literally a matter of life or death. However, the desire for energy services is also shaped in significant ways by social and cultural factors, because of the way energy is deeply bound up with the conduct of everyday life: as a result, what is deemed sufficient to cover basic energy needs in one social context may be insufficient in another. This chapter has highlighted the need for moving towards an integrated understanding of fuel and energy poverty. It has emphasised the socio-technical conditions that can create an insufficiency of energy services in the home, and the analytical value of approaching energy systems from the perspective of demand. We have also shown that individuals and groups across the world address the challenge of energy poverty and vulnerability in diverse ways, and highlighted the range of policies promoted in response to energy deprivation.

QUESTIONS FOR DISCUSSION

- What factors underpin geographies of energy poverty at the global scale?

- In what ways do young people in the global North experience energy/fuel poverty?

- Discuss whether strategies to extend energy access in off-grid areas of the global South necessarily address the wider dimensions of energy poverty.

- Explore the extent to which investing in green energy can complement efforts to address energy poverty.

- What are the advantages and shortcomings of the energy ladder and stacking models?

ACTIVITIES/POTENTIAL RESEARCH PROJECTS

- Examine available secondary evidence about energy or fuel poverty in a given country, and outline the causes and consequences of the problem

- Choose an individual country participating within the UN Energy4All initiative, outline its targets and the strategies in place for meeting them and consider the factors likely to affect their achievement.

RECOMMENDED READING

- Bouzarovski, S. and S. Petrova. 2015. A global perspective on domestic energy deprivation: overcoming the energy poverty–fuel poverty binary. *Energy Research & Social Science* 10: 31–40.

This paper argues that energy poverty is a global phenomenon and needs to be studied through a unified framework. It criticises the dichotomy of global

North–global South that frames past understandings of the issues. The paper proposes energy services and energy vulnerability as pathways towards the development of a planetary sensibility.

■ Liddell, C. and C. Morris. 2010. Fuel poverty and human health: a review of recent evidence. *Energy Policy* 38(6): 2987–97.

One of the most authoritative and comprehensive reviews on the relationship between fuel poverty and health. It draws on a number of large-scale studies undertaken over a period of ten years. One of its key innovations is the emphasis on mental health as opposed to only physical health. The paper is primarily focused on the global North.

■ Pachauri, S. and N. Rao. 2013. Gender impacts and determinants of energy poverty: are we asking the right questions? *Current Opinion in Environmental Sustainability* 5(2): 205–215.

This paper explores what we know about how women's well-being relates to energy poverty. It details the considerable gaps that remain in our understanding of those factors that impact upon women's experience of modern energy services and their decision-making power in relation to their adoption. It also investigates how existing gender inequalities can serve to undermine the potential for current energy access programmes to help transform women's lives for the better.

■ Sovacool, B. 2012. The political economy of energy poverty: a review of key challenges. *Energy for Sustainable Development* 16: 272–282.

This paper provides a comprehensive consideration of two of the debates touched on in this chapter – those concerning the concepts of energy poverty and energy ladders. In particular, it explores how energy poverty intersects with other facets of development (for example, those concerning equity, gender, health and environmental degradation). It also reflects on how energy poverty relates to key elements of the contemporary global energy system.

REFERENCES

Batchelor, Simon. 2015. Solar electric cooking in Africa in 2020 a synthesis of the possibilities. Available online at www.evidenceondemand.info/solar-electric-cooking-in-africa-in-2020-a-synthesis-of-the-possibilities.

Bhattacharyya, S.C. 2006. Renewable energies and the poor: niche or nexus? *Energy Policy* 34(6): 659–663.

Boardman, B. 1991. *Fuel poverty: from cold homes to affordable warmth.* London: Bellhaven.

Bouzarovski, S. and S. Petrova. 2015. A global perspective on domestic energy deprivation: overcoming the energy poverty–fuel poverty binary. *Energy Research & Social Science* 10: 31–40.

Buzar, S. 2007. *Energy poverty in Eastern Europe: hidden geographies of deprivation.* Aldershot: Ashgate.

Campbell, R., 1993. Fuel poverty and government response. *Social Policy & Administration,* 27(1): 58–70.

Haas, R., N. Nakicenovic, A. Ajanovic, T. Faber, L. Kranzl, A. Müller and G. Resch. 2008. Towards sustainability of energy systems: a primer on how to apply the concept of energy services to identify necessary trends and policies. *Energy Policy* 36: 4012–4021.

Harrison, C. 2013. The historical–geographical construction of power: electricity in eastern North Carolina. *Local Environment* 18: 469–86.

International Energy Agency 2011. *World energy outlook.* Paris.

Lee, K., E. Brewer, C. Christiano, F. Meyo, E. Miguel, M. Podolsky, J. Rosa and C. Wolfram. 2016. Electrification for 'under grid' households in rural Kenya. *Development Engineering* 1: 26–35.

Liddell, C. and C. Morris. 2010. Fuel poverty and human health: a review of recent evidence. *Energy Policy* 38(6): 2987–97.

Mahama, A. 2012. 2012 International year for sustainable energy for all: African frontrunnership in rural electrification. *Energy Policy* 48: 76–82.

Mekonnen, A. and G. Köhlin. 2008. Determinants of household fuel choice in major cities in ethiopia, environment for development. Discussion Paper Series. Available online at www.rff.org/files/sharepoint/WorkImages/Download/EfD-DP-08-18.pdf

Modi, V., S. McDade, D. Lallement, and J. Saghir. 2005. *Energy services for the Millennium Development Goals.* Washington DC: The International Bank for Reconstruction and Development/The World Bank/ESMAP. Available online at www.cabdirect.org/abstracts/20083156488.html.

Mortimer, K., C.B. Ndamala, A.W. Naunje , J. Malava, C. Katundu, W. Weston, D. Havens, D. Pope, N.G. Bruce, M. Nyirenda, and D. Wang, D., 2017.. A cleaner burning biomass-fuelled cookstove intervention to prevent pneumonia in children under five years old in rural Malawi (the Cooking and Pneumonia Study): a cluster randomised controlled trial. *The Lancet* 389(10065): 167-175.

Pachauri, S. and N. Rao. 2013. Gender impacts and determinants of energy poverty: are we asking the right questions? *Current Opinion in Environmental Sustainability* 5: 205–15.

Petrova, S., M. Gentile, I. Henrik Mäkinen and S. Bouzarovski. 2013. Perceptions of thermal comfort and housing quality: exploring the microgeographies of energy poverty in Stakhanov, Ukraine. *Environment and Planning* A 45: 1240–1257.

Rahman, S.M. and M.M. Ahmad. 2013. Solar Home System (SHS) in rural Bangladesh: ornamentation or fact of development? *Energy Policy* 63: 348–354.

Silver, M. 2015. If you shouldn't call it the Third World, what should you call it? Available online at www.npr.org/sections/goatsandsoda/2015/01/04/372684438/if-you-shouldnt-call-it-the-third-world-what-should-you-call-it

Sovacool, B. 2012. The political economy of energy poverty: a review of key challenges. *Energy for Sustainable Development* 16: 272–82.

Standal, K. 2016. The globalising effects of solar energy access on family and gender relations in rural India. *Asia in Focus,* Autumn 2016. Available online at www.asiainfocus.dk/

Tirado Herrero, S. and D. Ürge-Vorsatz. 2012. Trapped in the heat: a post-communist type of fuel poverty. *Energy Policy* 49: 60–68.

Uejio, C., O. Wilhelmi, J. Golden, D. Mills, S. Gulino and J. Samenow. 2011. Intra-urban societal vulnerability to extreme heat: the role of heat exposure and the built environment, socioeconomics, and neighborhood stability. *Health & Place* 17: 498–507.

Urmee, T., D. Harries and H. Holtorf. 2016. *Photovoltaics for rural electrification in developing countries: a road map.* Switzerland: Springer.

Energy consumption, inefficiency and excess

6

LEARNING OUTCOMES

- Understand the complexities involved in revealing patterns of energy consumption and associated carbon emissions.
- Appreciate the distinction between competing explanations for trends and patterns of energy demand, for example between those that focus on relationships between macro-level variables and those that emphasise the embedding of energy in everyday life.
- Explore the major features of alternative approaches towards reducing energy demand and apply critiques of the dominant energy efficiency paradigm.

In the end, all of the resource extraction, energy production technology and distribution infrastructure that makes up a big part of what we see as 'energy systems' have the purpose of enabling energy consumption – the charging of a mobile phone, the turning over of a car engine, the flashing on and off of a neon sign, the powering of a factory production line, the heating of a school. Even if we recognise that energy systems may have other organising logics that sustain them, the powering of end uses and the consumption that goes with this is still fundamental to their purpose and existence.

Energy use has become functional, expected and embedded in modern technologically driven societies in a multitude of different ways. Historically, as discussed in **Chapter 2**, total energy consumption aggregated to national or regional scales has been closely related to trends and patterns in economic growth. More energy availability and more energy demand have therefore been seen as a good thing, an indicator of progress and advancement. However, as made clear in many other parts of this book, there are big problems with such a simple equation. Energy production and consumption are deeply implicated in a host of major environmental problems and risks to health, and in geopolitical struggles and conflicts around the world. Furthermore, as made clear in **Chapter 5**, while securing access to energy services is important to well-being and quality of life, there are severe and problematic inequalities in patterns of access and affordability, which are often masked by aggregate figures of consumption. From these perspectives, therefore, more energy consumption is not simply a good thing that can further enhance living standards. Indeed, consumption is increasingly recast as being at the root of many of the global scale problems and challenges with which this book is concerned. Urry (2010), following this line of analysis, argues that unprecedented levels of energy use are central to the 'dark legacy' of the carbon society that is only just beginning to exact its effects.

The aim of this chapter is to provide a critical understanding of energy consumption and the demand that underpins it, considering ways in which energy use can be seen as excessive, wasteful and caught up in unsustainable dynamics of accelerating social change, commodification and globalisation. As we shall see, approaching energy consumption critically entails considering carefully what energy is for – and, therefore, how energy demand is produced and reproduced – rather than simply taking it for granted as an outcome of economic and social development. It also entails problematising the relationship between energy supply and energy demand, recognising the ways in which energy supply and infrastructural systems create and sustain energy demand in order to establish and maintain their purpose and productivity. Supply is not then just about satisfying what is demanded.

The first part of the chapter considers a set of issues around how we measure and track patterns of energy consumption and how they change. It makes clear that choices are involved in how energy and carbon data are represented, and the patterns that are thereby revealed. The second part examines different ways of conceptualising the making of energy demand and how and why it changes. It emphasises, in particular, an approach that sees energy demand as embedded in everyday practices. The third part of the chapter considers how notions of (in)efficiency and waste have been dominant in the evaluation of energy consumption. It shows how this has led to policies that seek to improve energy efficiency through technical innovation, incentives and standard setting, and eliminate what is seen to constitute the 'waste' of energy. The final part considers the limitations of focusing on efficiency alone, and the more fundamental challenges that can be made to current levels of energy consumption and the dynamics of demand which underpin them.

PATTERNS AND DYNAMICS OF ENERGY CONSUMPTION: DIFFERENT PERSPECTIVES

There is no doubt that, over the history of human development, there have been radical changes in how energy figures in everyday life. We can look to quantitative measures to get a sense of this transformation, but need to be aware that when attempting to measure something as distributed and differentiated as 'energy use', the unit of measurement, the type of indicator and exactly what is counted, all matter to the representation that is produced.

Looking back over long timescales there is no reliable data that can provide information on how much energy, in aggregate, has been consumed, at different periods of time. Analysts have to produce estimates based on what data and assumptions they can meaningfully put together. Fouquet and Pearson (1998), for example, attempt to track changes across a thousand years of energy use in the UK by bringing together a patchwork of partial and incomplete historical data on patterns of economic activity, prices of various forms of commodity, population dynamics and historical shifts in how energy services have been derived through different forms of technology. Sørensen (2012), even more ambitiously, has produced a 100,000-year analysis of energy use in Northern Europe, from Neanderthal society onwards. This necessarily includes periods in which very sporadic archaeological records have to be relied on, along with inferring the energy needs of humans through modelling of climatological conditions and other indirect indicators.

The focus of more recent data has been on commodified and traded energy, whereby energy consumption statistics are derived from evidence on transactions of selling and buying energy sources and from grid monitoring systems. Despite the existence of these indicators, where energy consumption involves wood and biomass that is informally collected, where local trading systems are used, or where energy generation is 'off-grid', it can often very difficult for all instances of consumption to be known and meaningfully aggregated (adding, for example, to the complexities involved in tracking the progress being made against international energy access targets: Bhatia and Angelou 2015).

The major global databases – produced by the United Nations, the International Energy Agency and the BP Statistical Review of World Energy – go back only to around the 1950s and within their earlier periods are often very incomplete. As shown in Table 6.1, in these databases different measures are used, giving different indicators not just of absolute energy consumption but also the percentage change in consumption over time. There are evidently options available here in terms of which representation of the scale of change in energy consumption is to be chosen, rather than simple 'facts' to be reproduced.

Table 6.1 Alternative measures of change in global energy consumption

Source	Global measure	1973	2012	% change
International Energy Agency	Total final energy consumption	4661 mtoe*	9117 mtoe	196
BP Statistical Review of World Energy	Total primary energy consumption	5692 mtoe	12633 mtoe	221

Notes: *mtoe = million tonnes oil equivalent. Total primary energy consumption includes the energy transformed within supply systems e.g. in using gas to generate electricity. Total final energy consumption excludes such conversion and only estimates final end use outside of the energy system.

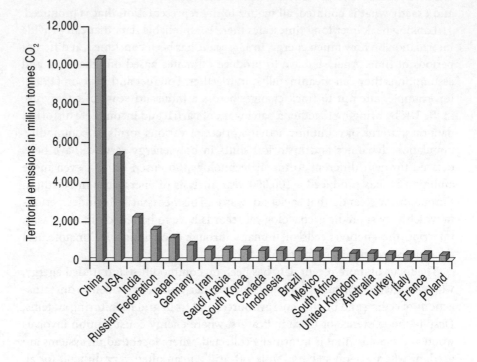

Figure 6.1 Total carbon emissions by country in 2016

Such totals are routinely broken down into groups of countries, individual countries, sectors (industrial, transport, domestic), but also into per capita figures that allocate an equal proportion of this consumption to each citizen of the geographic unit involved. Per capita figures can provide very different impressions to absolute ones, and have been particularly influential – and controversial – in debates about carbon emissions. As shown in Figures 6.1 and 6.2, the responsibility for carbon emissions between countries looks quite different depending on whether absolute or per capita figures are considered, with China, in particular, standing out as having high total emissions but relatively low per capita ones. Our perspective can change again, however, if total historical emissions accumulating over past decades or centuries are taken into account rather than only contemporary emissions. Choosing the indicator to focus on is not just a technical matter, and has been highly controversial in the politics surrounding international climate change negotiations (Friman and Hjerpe 2015).

While some people have argued on justice grounds that per capita data are most appropriate as a guide to the allocation of emission targets between countries (Meyer 2000), this statistical way of dividing up emissions can in turn be critiqued as ignoring variation *within* country units. In reality, people are not equally responsible for carbon emissions in terms of the energy they

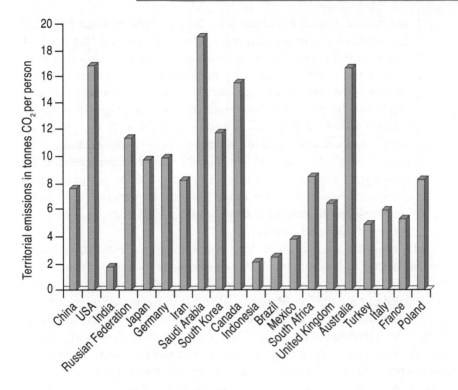

Figure 6.2 Per capita carbon emissions by country in 2016

actually consume at home, in moving around and so on. Studies have shown, for example, that in the UK energy consumption and carbon emissions vary significantly between different households, following lines of class and income differences. Those on higher incomes have much higher energy consumption footprints than those on low incomes (Brand and Boardman 2008). Preston et al. (2013), in an analysis of energy use in the home and for personal travel, calculated that the richest 10% of households in Great Britain emit three times the carbon of the poorest 10%. Even wider disparities can exist in the global South, where average per capita figures are low, but elite and wealthy parts of society can have lifestyles that are very carbon intensive. It is also questionable whether, as is the case with some datasets, it is reasonable to allocate emissions from industry, government, retail and similar activities on a per capita basis, making individuals appear to be in some way responsible for all forms of emission, not just those related more directly to their everyday lives.

We can get a different picture again by shifting the boundaries within which calculations are made. The notion of **embodied energy** highlights that all items that are produced, transported and sold require energy consumption during their production, transportation and selling. The

embodied energy
the energy consumed by all of the processes associated with the production and provision of any object or commodity. For example, for a building this would include the energy used in the mining, transport and processing of the materials that the building is constructed from, such as bricks, metal beams and concrete.

offshoring
the relocation of business
activities from one country
to another, often from
countries where there are
higher business costs and
tighter standards and
regulations, to ones in which
costs are lower and
regulations less strict.

significance of such calculations for how we view patterns of energy consumption and responsibility for carbon emissions is that much of what is consumed in more wealthy countries is now produced in other parts of the world. Decades of progressive shifts in patterns of production – the outsourcing and **offshoring** that have transformed geographies of the global economy – mean that there are major dislocations between spaces of production and consumption. It then becomes pertinent to ask where and to whom the energy consumed and carbon emitted should be allocated. Does it stay with the country doing the producing – for example, Taiwan a country with an economy based strongly around making consumer products for export – or with the countries, and the citizens of those countries, doing the purchasing and consuming of those products?

Again, big shifts are possible in how problems are represented, as studies have shown that 20–25% of global CO_2 emissions are from the production of internationally traded products (Davis and Caldeira, 2010). Figure 6.3 from a study carried out for the UK (Barrett et al. 2013) shows trends in greenhouse gas emissions under three different boundary conditions – 'territorial-based', which is the conventional way of assigning emissions to the land and offshore territory of the country that they are physically associated with; 'production-based', which adds in international air and shipping travel that is beyond any national territory, assigning it to the country of the travel operator or the

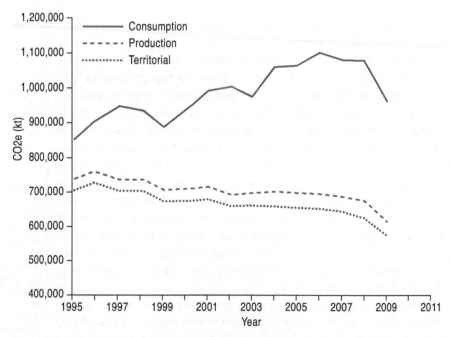

Figure 6.3 UK greenhouse gas emissions differentiated between consumption, production and territorial methods of calculation

EXAMPLE 1: THE MARKETING OF SUVS

In the 1990s a major controversy erupted over the environmental implications of increasingly popular Sports Utility Vehicles (SUVs), the advertising campaigns that promoted them and the lifestyles they were said to represent. Vehicle manufacturers promote SUVs because they provide significantly higher profit margins than compact cars and sedans. SUVs evaded federal fuel economy standards for passenger vehicles in the US market: based on the chassis of a pick-up truck or commercial vehicle, SUVs were classed for regulatory purposes as 'light trucks'. Given their poor levels of fuel efficiency (studies at the time suggested that driving an SUV produced 8,000 lbs (3,600kg) more CO_2 than driving a regular car over the course of a year), those who drove SUVs were presented as embodying thoughtless energy-guzzling consumerism and environmental illiteracy. Drivers became the subject of direct action protests, ranging from leafleting campaigns to arson attacks on SUV dealerships. Some of the ire directed at SUVs stemmed from the incongruity of their woeful environmental performance with marketing campaigns that celebrated an SUV's capacity to facilitate access to unspoilt nature and wilderness. As Aronczyk (2005) remarks, "even the names of these vehicles – Land Rover's *Discovery* and *Freelander*, Subaru's *Forester* and *Outback*, Ford's *Excursion*, *Expedition* and *Escape* – underline the motorist's journey as one of an intimate connection with the natural environment". As time has gone by, the SUV has to some degree fallen off the political radar as manufacturers have made innovations that have been designed to address at least some of the worst features of the earlier models. Ford, for example, brought in hybrid versions of their SUVs relatively early, there are new electric SUVs appearing on the market and the fuel economy of traditional SUVs has improved considerably. High oil prices in the early 2000s dented SUV sales but they subsequently rebounded, and today SUVs represent the largest segment of the global car market. Despite the earlier controversy, the auto-industry continues to emphasize the capacity of SUVs to access unspoilt nature on demand – and without any trace of irony. If anything, contemporary advertising ups the ante, constructing SUV drivers as environmentally conscious agents exercising responsible choices: adverts promote SUV drivers as deliberately protecting nature, manufacturers sponsor environment-themed events, environmental slogans adorn tyre covers, and so on. So, despite SUVs generally still being amongst the most poorly performing vehicles in terms of fuel efficiency, they are expressly marketed to those who want to be seen as environmentally conscious. Rollins (2006: 711) summarises the SUV phenomenon effectively,

> It appeals to many consumers' deeply felt desire to engage with their natural world, and to a certain extent gives evidence of the greatly increased environmental awareness that is characteristic of the postmodern age. What makes the SUV so outrageous is that it is a twisted expression of that developing environmental consciousness, a perversion of energies that might, collectively, have built something far more sustainable by now.

residence of the tourist traveller; and 'consumption-based', in which emissions are allocated to the country of the end consumer. The differences in the trends reported for the UK are striking with 'territorial' and 'production' emission indicators declining from 1990–2009, while the 'consumption-based' measure heads strongly in the opposite direction.

It is clear then that it is important to think critically about energy consumption (and carbon emission) data, how it is produced and what it is representing. There is no doubt that energy consumption globally has grown historically, and that it continues on an upward trend. But beyond the global picture patterns of change are very differentiated and can be viewed in many different ways. In this respect, claims made about future energy demand can be politically contentious, particularly where they are being used to justify the expansion of controversial energy supply projects (Walker 2016).

UNDERSTANDING THE DYNAMICS OF ENERGY DEMAND

While the types of data discussed in the previous section are broadly instructive, data alone have little explanatory power. We can get a sense of where consumption (or emissions) are higher or lower (by different measures), what 'sectors' are growing or contracting and what particular 'fuels' are being consumed in simple terms, but cannot in any sense *explain* the patterns that are shown.

Many explanatory frameworks of the relations between energy use and societal change work at a macro level (Shove and Walker 2014), focused on the interrelations between 'factors' understood in general terms, such as technological innovation, energy costs, population growth and economic development and change. Studies looking back historically have worked with rather linear models, focused on transitions in changes in dominant fuels (Hirsh and Jones 2014) and seeing patterns of cause and effect in simple terms. A striking example is White's (1943: 338) classic study of energy and the 'evolution of culture' in which he states with a law like confidence that:

> culture develops when the amount of energy harnessed by man per capita per year is increased; or as the efficiency of the technological means of putting this energy to work is increased; or, as both factors are simultaneously increased.

Such deterministic (and gendered) language is less used today, but even so, energy consumption can still be presented as wrapped up in processes of change that have an inevitability about them. As economies and populations grow, as new technologies emerge and diffuse smoothly and predictably across space, as urbanisation and globalisation progress, it is assumed that energy demand will respond and increase accordingly.

Various challenges have been made to such simplistic understandings. Hirsh and Jones (2014), for example, take issue with assumptions of continuity with historical experience, arguing that analogies between the past and present can be misleading in terms of the apparent tight and consistent relationships between macro variables that they are taken to imply. Cultural variation, Wilhite (2008) argues, is particularly absent from such thinking. More complex understandings of processes of innovation and transition have also been developed, which emphasise the different speeds and scales of change involved and the ways in which certain energy pathways can become locked-in and others resisted (Araújo 2014) – see **Chapters 9 and 10** for further discussion.

While technological change is undoubtedly important in producing new forms of device through which energy is used, as emphasised in **Chapter 3** and elsewhere in this book, technology is always embedded in social processes rather than independent of them. This way of thinking is fundamental for Shove and Walker (2014) who argue that the dynamics of energy demand can properly be understood only through focusing on what energy is for. Demand, they argue, comes from social practices, the everyday activities through which energy is used in order to achieve valued outcomes – eating cooked food, travelling to shops to make purchases, reading or doing homework in the evening, communicating with others via the web and many other such normal and socially shared aspects of everyday life (see **Example 1**). They conceive, therefore, of energy demand as thoroughly embedded within the ongoing reproduction and transformation of society, arguing that:

> understanding trends and patterns in energy demand (and in provision and supply as well) is in essence a matter of understanding how social practices develop, change and intersect.
>
> (ibid.: 47)

The "in provision and supply as well" in this extract points to an important part of their argument, which is that the arrangements through which energy is supplied and distributed – the infrastructures discussed in **Chapter 3** – are directly implicated in the making and ongoing reproduction of energy demand. Infrastructures make energy demand possible and create dependencies in everyday life on the continued enactment of energy-consuming practices. As argued in **Chapter 3**, infrastructures, therefore, are involved in the making of demand, rather than in simply supplying it. This can be seen, for example, in emerging literature on the gradual growth in power demanded amongst customers of contemporary off-grid energy developments in the global South once households achieve initial access to electricity (Scott and Miller 2016).

To illustrate these ideas further we can consider the case of indoor air conditioning (Figure 6.4). Starting in America in the early twentieth century (Cooper 1998; Ackermann 2002), air conditioning has spread around the world, increasing energy consumption and producing energy demand where none existed before. The continued global growth of air conditioning as an increas-

Figure 6.4 Domestic air conditioning, Taiwan (photo credit: Gordon Walker)

ingly normal feature of building design and everyday living is seen as a major threat to demand and emission reduction objectives, particularly if the density of air conditioning that is standard in the US is reproduced in other parts of the world (Sivak 2009). However, the evolution of air-conditioned space has played out in different ways geographically, with, for example, studies of experiences in different parts of Asia pointing to both patterns of similarity and distinctiveness (Wilhite 2008; Sahakian 2011; Hitchings and Lee 2008). Drawing on such evidence, Shove et al. (2014) explain the rise of air conditioning not as a simple story of technology diffusion associated with increased prosperity, but as an outcome of many multi-sited processes in which air conditioning has become integrated into practices like those of office work (Hitchings 2011), nursing, taking a luxury holiday, driving in warmer weather, or being at home. Air conditioning is often in the background for these practices but part of how they are now seen to be carried out.

Taking a case study of a specific hospital in the UK, Walker et al. (2014) use this approach to explain the surprisingly rapid and recent growth in the number of air conditioning units at the hospital. They conclude that the spread of air conditioning was an outcome of repeated moments at which multiple, situated forms of apparent 'need' become established within particular buildings, closely related to how these buildings were being used (the practices they were the site for) and how those practices were changing. The increasingly intensive use of spaces within hospital buildings, the introduction of new heat producing and heat sensitive technologies as part of nursing, communicating and catering, and the increasingly 'risk averse' culture of hospital management, were all implicated in the detailed explanations they draw out.

Another good example of how energy use is embedded in the dynamics of everyday practice comes from the rapid changes in the use of information and communication technologies (ICT). Keeping track of friends, booking train travel, doing work on the move, cooking meals from online recipes, trading shares, watching rugby matches, to take just a few examples, have all become practices that now make use of ICT of various forms in a way that they didn't before. Røpke et al. (2010: 1764) characterise the dramatic rise of ICT as a "new round of household electrification" in which the use of multiple devices, sometimes simultaneously, in new patterns of watching, communicating, searching, ordering and scheduling has become entirely normal and pervasive amongst Northern societies (and increasingly, although far from universally, in many countries of the global South: World Economic Forum 2015). This has implications not just for the direct consumption of energy by the IT devices themselves, but also for the extended infrastructure of the internet, servers, cloud storage and similar, through which these IT devices are digitally connected (Wiig 2017. Every byte of transmitted and stored data is a demand for energy (including air conditioning to keep vast server rooms distributed around the world at operational temperatures – for example, see the Case study of digital data's energy landscapes in Norrbotten, Sweden in **Chapter 2**).

Conceptualising energy demand in terms of the relationship between practices and infrastructures has also been important in developing insights into the temporal patterns of energy demand – in particular, the peaks and troughs of electricity use that are increasingly problematised within energy policy (see **Chapter 3**). A practice approach to analysing the daily, weekly and seasonal dynamics of energy demand emphasises that these are made by patterns of social synchronisation, people doing similar things at similar times. The rhythms of society as a whole can thus be observed at an aggregate level, but they are made up or constituted by the many practices of people and organisations reproducing, over time, similar patterns of coordinated activity. As Walker (2014: 52) describes:

> The classic aggregate temporal pattern of electricity demand over a typical weekday, has a morning peak and an evening peak. These recurrent peaks are produced through the social synchronisation of energy demanding practices (across multiple spaces) into the same time periods. This might

mean the synchronisation of the same practice (e.g. cooking) into the same peak time period, or the bundling of multiple interconnected practices into the same peak period. For example, cooking, watching TV, using computers, having lights on, running dishwashers, doing the vacuuming, all happening during the evening peak when there is much home based activity. Each separate practice is not necessarily precisely synchronised (everyone washing the dishes at the same time) but there is a shared pattern of energy consuming practices of various forms happening during the same early evening peak period – a shared pattern that might be integral to the understanding of the practice itself as well as to its temporal positioning in relation to other practices.

This type of understanding of how energy demand is built into the temporal structures of society has implications then for attempts to reduce the intensity of peaks, to 'load shift' and make demand more responsive to peak period pricing and similar 'smart' initiatives (Torriti 2016). Different forms of implication can also emerge in the context of the inefficient and frequently overloaded national grid systems of many states in the global South. Fear of 'load shedding' (the temporary turning off of sections of the grid) has been, for example, a major factor influencing the lack of take-up of electric cooking devices and the limited promotion of electric cooking as a truly 'clean' alternative to the burning of biomass across Sub-Saharan Africa (with the partial exception of South Africa: Brown et al. forthcoming).

THE PROBLEM AS INEFFICIENCY AND WASTE

As noted earlier, a key reason for wanting to understand more about energy demand and what it is made up of is because of arguments that we should be taking measures to reduce it. This objective can be fitted with many policy concerns – enhancing energy security, avoiding investment in new production and generation, reducing pollution and carbon emissions, reducing business costs and addressing fuel poverty – making it, for some people, the most important and common sense focus for contemporary energy strategies.

energy efficiency is generally defined as an objective of delivering the same energy service or outcome with less energy input. For example, heating a room to a given temperature with less use of energy, by improving insulation and draught proofing or installing a new more efficient heating system.

By far the dominant way of approaching energy demand reduction is to focus on improving **energy efficiency** and reducing waste. Indeed, often demand reduction and action on energy efficiency are seen as entirely synonymous. Various other terms have also been deployed including 'energy saving' and 'energy conservation' and while technical distinctions can be drawn between them they are all in essence getting at the idea of achieving the same ends but with less consumption of energy in the process (an idea returned to later).

Efficiency more broadly is a notion that Lutzenhiser (2014: 143) relates to "modernity in general and . . . more deeply to a protestant worldview" with a substantial energy efficiency industry emerging around these ideas since the 1970s and the first energy crisis. He characterises this industry as encompassing "the coordinated actions of utility companies, government agencies, business firms and non-profit advocacy groups in the process of producing energy efficiency as an output" (p.142). This output is rather strange, he adds, as it an absence of something that did not occur – energy flows that did not happen – rather than a tangible object or service.

Certainly, estimates of what can be achieved by being more energy efficient have proliferated, and can be found attached to particular technologies, individual homes, whole housing stocks, to cities, industries, economies and ultimately the whole global energy system. Some examples are listed in Table 6.2. Often, the drive behind such estimates is to point to the vast unrealised potential there is in 'saving' energy rather than producing and consuming it. Being more efficient has in these terms been framed as an alternative to supply, something that can be invested that achieves a better rate of return than building new power stations. The notion of **negawatts** (Lovins 1990) captures this logic – avoiding megawatts of production by instead achieving negawatts of efficiency gains.

Many different types of policy measures have been developed and implemented in order to try and promote greater energy efficiency.

negawatts
a way of expressing the gains that can be achieved by energy saving. Negative watts are units of electricity consumption that can be avoided by investment in energy efficiency and reducing waste.

Table 6.2 Examples of energy efficiency policies and proposals

Energy efficiency policy	Potential to improve efficiency
EU Policy on Improving the Energy Efficiency of Buildings	The European Commission suggest that buildings account for 40% of energy consumption and 36% of CO_2 emissions in the EU. This is mainly related to heating; they argue that while "new buildings generally need fewer than three to five litres of heating oil per square metres per year, older buildings consume about 25 litres on average. Some buildings even require up to 60 litres". They estimate that around a third of the buildings across the EU are at least 50 years old and that by "improving the energy efficiency of buildings, we could reduce total EU energy consumption by 5–6% and lower CO_2 emissions by about 5%". Source: https://ec.europa.eu/energy/en/topics/energy-efficiency/buildings
UN SEforALL Initiative	The UN's SEforALL initiative seeks to bring about universal access to modern energy services by 2030. In order to ensure that meeting such an ambitious goal does not contradict global emission reduction objectives, this commitment is accompanied by two other goals: doubling the share of renewable energy in the global energy mix and doubling the global rate of improvement of energy efficiency. The Energy Efficiency Committee Report to the Advisory Board of SEforALL in 2014 estimated that this would require global "investment in energy efficiency measures to rise by approximately $130 billion per year over the recent level of $180 billion per year in order to reach the SEforALL objective".
UK National Energy Efficiency Strategy	In 2012 the then Department of Energy and Climate Change launched a national energy efficiency strategy that it suggested could cut UK energy use by 11% by 2020. It focused on tackling four major obstacles to improving energy efficiency: market imperfections, lack of information, weak incentivisation and practical difficulties in installing efficiency measures. The report argued that its proposed programme of investment in tackling these barriers could save the UK 196TWh by 2020, equivalent to the output from 22 power stations.

These range from providing information on energy-saving potential, specifying technical efficiency standards (on appliances, houses, cars etc.), investing in research and development, banning inefficient technologies, using taxes and subsidies to incentivise economic gains and sometimes providing free or subsidised installation of more efficient devices. Lutzenhiser (2014) sees these established measures as all falling within the frame of a 'physical-technical-economic model' (PTEM) which he characterises as:

rational actors
a term related to the widespread assumption made within policy interventions that people will follow clear economic and self-interested logics in making choices between alternatives. This has been much criticised as an inadequate and very reductive understanding of what determines human behaviour.

- focusing almost exclusively on devices and costs, with little consideration of social systems, consumers as social actors or other non-technical and non-economic dimensions;
- monetised costs and benefits of investment in energy efficiency are of primary importance; cost effectiveness is required with an economic return;
- consumers are rational actors who are intrinsically interested in economic costs and returns, they are information seeking, calculative and conscious of the purposes of their use of energy;
- consumers are typically considered through measures and assumptions based on averages, that indicate the unexceptional and normal way for them to behave.

Looking back, attempts to improve efficiency and remove waste following these logics some efficiency gains have certainly been achieved. For example, while the millions of dollars of funding that have gone into the clean cookstoves (see Figure 6.5) sector over recent years have generally been justified on the basis of their supposed beneficial improvements to health (although see Mortimer et al. 2017), the sector has also been promoted on the basis of the enhanced efficiency of such stoves (efficient stoves are said to be 30 to 60% more efficient in their use of biomass fuel than traditional stoves). The Global Clean Cookstoves Alliance progress report for 2016 suggests that "(c)umulatively, an estimated 82 million stoves and fuels, including 53 million clean and/or efficient, have been distributed since 2010".

However, the literature on energy efficiency is also full of discussion of barriers and obstacles, finding a wide range of problems in realising what is deemed to be technically achievable and apparently rational to do in economic terms – in addressing the 'energy efficiency gap'. Responses have then followed which, staying within the same way of thinking, have emphasised the need to communicate better, find ways to get people to behave more rationally, to remove market barriers getting in the way of innovation diffusion and develop more effective economic incentives (for an excellent discussion of these pitfalls in the case of clean cooking transitions see Crewe 1997).

More critical perspectives have though taken on the assumptions of the PTEM model framing, arguing that these are distanced from social realities, in particular in terms of assumptions about consumers as primarily rational actors, and technologies as separate from the social processes and contexts through

Figure 6.5 Improved cookstove and smoke hood, Nepal (photo credit: Practical Action/Prabin Gurung)

which they come to be deployed (or not deployed) (Lutzenhiser 2014; Shove 1998). A socio-technical perspective challenges such assumptions, focusing on how energy using devices are caught up in complex ways in the everyday working of people's and organisations' practices, how varied and often contradictory rationalities and framings are at work rather than just an economic one, that power to achieve change is unevenly distributed, and much else that serves to make the assumptions of the PTEM an 'energy efficiency wonderland' (Lutzenhiser 2014: 148). An increasing number of studies have taken up these ideas. The **Case study** provides an example explaining the limited use of sustainable energy technologies in care homes for older people.

Other perspectives have pointed to ways in which efficiency gains, even if implemented successfully, can be overwhelmed by other ongoing dynamics, meaning that the impact on aggregate energy demand is not as significant as it appears. The **Jevons paradox** and closely related 'rebound effect' (Chitnis et al. 2013, see also **Chapter 9**) have been used to capture the ways in which the money savings achieved through using less energy may end up being used on other energy using activities (for example, savings on petrol from driving a more energy-efficient car being used to drive that car further than it would have been). This way of thinking about dynamics stays very much within an economic framing of human action, but more sociological analyses go further to point to the many forms of ongoing social change that can be at work in parallel with progressive energy efficiency improvements. These can include, for example, shifts in population dynamics (more single-person

Jevons paradox
refers to the counter-intuitive way in which gains in efficiency via technological change end up expanding (rather than decreasing) resource consumption. The phenomenon was noted by the British economist William Stanley Jevons, who observed how the enhanced efficiency of steam engines lay behind the country's dramatically increasing consumption of coal. Efficiency gains reduce the cost of fuel as a portion of total cost, making it possible to consume more.

163

CASE STUDY: SUSTAINABLE THERMAL TECHNOLOGIES
IN CARE HOMES FOR OLDER PEOPLE

In a study of the implementation of sustainable thermal technologies in care homes for older people, Neven et al. (2015) demonstrate the complexities of integrating new technologies into specific settings, even where the evident gains to be made in terms of improved efficiency and savings on energy bills seem readily apparent. Care homes for older people have characteristics that in technical and economic terms suggest that they might be particularly appropriate for the implementation of more sustainable thermal technologies. They have comparatively high demands for space heating and hot water often sustained on a 24/7 basis, and providing a warm environment is seen, in the UK in particular, as central to caring for people in their old age. However, there are many considerations, both generic and contextual, that will typically play into processes of technology uptake and for care homes there are particular institutional qualities to consider.

Care homes are a hybrid category of building use and indoor environment – part long-term residence, part nursing space, part working environment – slipping between conventional categorisations of home and work and public and private space (Walker et al. 2015c). Care homes also come in very different forms, some in old buildings, some very new (and combinations of old and new). Some are part of large corporate chains, others are run by local authorities (although declining in number), or are individual homes run by owners/managers who live on-site. These qualities were important to the findings from research undertaken in six case study homes, which was focused on management and staff perspectives and experiences rather than on just technical questions.

Two key themes emerged. The first related to the balancing between the cost savings achievable through making new technology investments, and the risks to the business that going into unfamiliar technical territory was seen to entail. On the costs side, the savings that could be made were understood, but were often seen as marginal to the overall costs of running the care home, even despite rising energy prices. On the risk side, there was something of a mixed view, but mainly great sensitivity to anything that could lead to either disruption to care home operation (which have to run 24/7) or damage to their wider reputation as a care provider. As the researchers comment, the reputation of a care home could in this way be seen as dependent on something as seemingly mundane and obscure as the availability of parts to mend a broken ground source heat pump.

The second theme related to impacts on care practices for older and vulnerable residents, emphasising both their importance but also the benefits and sometimes dis-benefits that certain technological arrangements could bring, particularly those that facilitated more stable thermal environments. To use the example again of the ground source heat pump and underfloor heating, such an installation could be encouraged as much by the removal of radiators and the greater safety and flexibility that this provides for care practices, as by the energy and cost savings that are achieved. But negative consequences were also seen from having hard flooring and from issues with cleaning and odour management. Technologies are not just evaluated therefore in terms of their energy-saving potential.

Ultimately, the research concluded that if care home owners see the various potential risks to their operation as significant and not easily mediated, and the benefits as uncertain and/or insubstantial, they will unsurprisingly be conservative and stick with what they already know. Even more so when the opportunities for making infrastructural changes are so limited in buildings that are continually in use for the caring of vulnerable older residents. It could be, they conclude, that the care home sector remains rather reluctant to embrace sustainability innovations, despite the clear technical potential that is out there – an assessment that clearly indicates the complexities involved moving from technical potential to its material realisation.

households), in the patterns of use of technologies (e.g. more TVs in the home, switched on for more of the day) and in other features of technological design (e.g. larger and heavier cars) and their integration into everyday practices, which can serve to push total energy demand up thereby counteracting efficiency gains. In an analysis focused on Sweden, Bladh (2012: 386) comments that:

> There are . . . counteracting tendencies of a more purely social nature that partially outdo the gains in efficiency in the more purely technological sense. Increasing population, a higher number of cars per capita and fewer people in the average household are examples of such tendencies. They play a significant role in the total consumption of energy, probably more important than energy rebound effects.

Again, then, we can see the importance of approaching energy use as a social matter, deeply embedded in the ongoing dynamics of societal change rather than separable from them. The pursuit of improved energy and carbon efficiencies can also be approached through a political-economy lens. This would recognise, for example, how energy suppliers do not necessarily see it as in their interest to support reductions in consumption, that governments often generate significant tax revenues from energy production and consumption (e.g. fuel sales), and that companies charged with implementing regulatory standards are also sensitive to potential negative impacts on their profits. A recent example from the UK bears out such an analysis: a new **zero carbon** standard for all new housing in England, developed through a nine-year collaborative process, was dropped by the government in 2015 following lobbying from conservative voices in the housing industry (Walker et al. 2015a, 2015b). The sustained failure to realise energy-efficiency potentials, and neglect of the 'demand-side' in energy policy priorities, can therefore be seen as a reflection of the unequal distribution of political power across the wide spectrum of interested and involved actors.

zero carbon
an objective set for a particular defined space (such as a house, workplace or a town) or an activity such as a train journey. To meet this objective there will be no net addition of carbon to the atmosphere. This may be achieved by all of the energy used coming from renewable sources, and/or by an offsetting of carbon emissions through, for example, the planting of trees. Exactly how the 'zero' is calculated can be both complex and controversial.

THE PROBLEM AS CONSUMPTION AND DEMAND

The more critical arguments already discussed point towards both the inadequacies of established ways of conceptualising energy efficiency and the limitations of relying on enhancing efficiency as a way of reducing energy demand. In the context of the speed and extent of reductions in carbon emissions that many have called for, these limitations have been becoming all the more evident (Walker 2016). Another way of approaching demand reduction, then, is to problematise the basis of demand itself: the goal is not to make energy use more efficient, but address what energy is being used for in the first place. This means focusing on established norms of living in energy-intensive societies, as well as the actions of many commercial and governmental actors involved in promoting energy using forms of consumption and sustaining continued dynamics of energy-hungry economic growth (including that promoted to the most rapidly growing economies of the global South: see Rafey and Sovacool (2011) for a discussion of discourses of energy development in South Africa).

Moving into this territory involves questioning what is needed in society in order to achieve well-being or notions of the 'good life' (Sayer 2011), as well as arguing that responsibilities to reduce and control carbon emissions do have to be acted on (see **Example 2**). It is not simply a matter of saying that energy consumption should be reduced across the board, as there are deep inequalities in patterns of energy use both between and within countries and there is a clear need to bring energy access and better quality energy services into the lives of many people who do not have them (see **Chapter 2**). But who is to say where the boundary lines are between what is essential energy use, what is wasteful, what is unnecessary and what is excessive?

Walker et al. (2016) take on these questions in analysing a set of focus group discussions (repeated every two years from 2008 to 2014) involving members of the public and focusing on what constitutes a minimally decent living standard in the UK. They examine what energy-using items are included in the expected ways of living that these groups arrive at. Core energy uses such as for heating, lighting, cooking and entertainment are present throughout the group discussions across the whole time period. Others were added in as notions of what was normal in everyday life evolved over time – such as computers, larger-screen TVs and cars for bigger families. This shows how understandings of 'what is essential' do shift over time relative to the prevailing societal context and that there can be an ongoing escalation of norms of energy dependency and consumption.

Such debates over what might be considered essential in terms of the level of energy consumption needed to access well-being and a decent life are mirrored in discussions over the developmental impact of energy access projects. For example, there has been controversy over what constitutes 'access' to modern energy services and hence what will be measured in terms of monitoring the achievement of the Sustainable Energy for All (SEforALL) target of achieving universal energy access by 2030. As Todd Moss (2015) of the Center for Global Development puts it,

> The current definition of modern energy access used by the International Energy Agency, the United Nations, and reported by the World Bank uses a threshold of 500kWh per year per urban household – and just half this rate for rural households. This means an international definition of modern energy access at 50–100 kWh per person per year. That's what a typical American uses in less than three days. It is just 3% of the global average and roughly enough power to charge a cell phone and run a few lightbulbs for a few hours a day. That is unambiguously not modern.

This type of 'bottom line' approach has occasionally given rise to a somewhat patronising attitude amongst donor-funded projects where household systems have been engineered so that they will only provide lighting or potentially mobile phone charging but cannot (without 'tinkering') be used to run appliances like televisions or radios, which are seen as somehow frivolous and not associated with social improvement. Gent (2014: 178) explores this in relation to a project run by a major NGO in Nicaragua where an inverter enabling the

---

EXAMPLE 2: BUY NOTHING DAY

Buy Nothing Day (BND) originated in Canada in the early 1990s and has subsequently spread to over 60 countries as an international form of collective action against consumerism. Since 1997 BND has been organised to coincide with Black Friday, the promotional sale events at the end of November that follow the US Thanksgiving holiday. Here BND serves as a direct rebuttal of the efforts of retailers and advertisers to accelerate and expand consumption. In countries where Black Friday is not promoted, BND takes place on the following Saturday.

The primary focus of BND is to encourage individuals to make a personal commitment not to purchase anything on one particular day. This has been the source of some criticism, as arguably all it does is to move consumption to a different day. However, BND's wider intention is to call into question norms, desires and behaviours around shopping and consumption, and raise understanding of the factors influencing our decisions about what we (want to) buy. To this end, BND is supported and enabled by a range of organisations such as the Vancouver-based Adbusters, a global network of artists, writers, musicians and designers who seek to unsettle, subvert and transform established consumption norms. As the Adbusters website (2017) puts it,

> Buy Nothing Day isn't just about changing your habits for one day. It is about rediscovering what it means to live freely. If you are going to buy, make a choice and go local, independent, or make something. Let's take back our lives and stop buying into the consumerist machine.

The ultimate aim is to nurture "an economy and a culture in which it eventually becomes cool to consume less" (Revkin 2011).

---

use of AC devices was not included in their SHS package because the organisation did not consider financing the facilitation of the use of TVs to be an appropriate use of international development aid monies. As might be imagined, in most cases local community members had adapted the systems to be able to use the systems for such purposes.

This reflects the differing assumptions held by donors, project developers and end users over quite what the role of energy provision should be in meeting the desires and aspirations held by poor communities. Gent (2014: 177) draws on Escobar (1995) to argue that the current drive towards meeting universalised energy 'needs' (however those needs are defined) can serve to render "people's needs as unproblematic" when, in reality, individual communities exhibit different desires, priorities and aspirations in relation to energy use. As she argues, this discourse can provide "legitimacy for intervention in the form of aid or technical assistance" imposing "external (Northern) norms and expectations onto other societies" that reject other ways of living as inferior.

degrowth

a new political and social imagination that has emerged in opposition to the ideology of economic growth, which continues to dominate contemporary policy discourse. It has its roots in the *décroissance* movement in France, although over recent years it has been growing in significance more widely across Europe and the Americas. It is sometimes contrasted with other connected notions such as a-growth, post-growth or anti-growth.

This brings us back to a consideration of more political arguments about energy demand that are based not so much around improving the efficiency of how current levels of energy consumption are delivered, but rather questioning the basis of that demand itself by challenging the growth-fixated dynamics of modern political and economic discourse globally. Recent years have seen the emergence of a range of social movements concerned with both challenging the dominant political and economic forces that continue to promote current patterns of energy demand, and presenting alternative ways of living that to a greater or lesser degree involve some kind of rejection of those patterns. One increasingly influential way of conceptualising these alternatives comes from the idea of **degrowth** which, as Martinez-Alier et al. (2010: 1741) put it, is emerging as "both a banner associated with social and environmental movements and an emergent concept in academic and intellectual circles". Amongst activists, degrowth and associated concepts such as a-growth (Jeroen et al. 2012) have helped to solidify an oppositional utopianism or 'confluence point' for several streams of critical ideas and political action around a hitherto disparate range of seemingly localised struggles, particularly across Europe (D'Alisa et al. 2013; Kallis and March 2015).

CONCLUSION

This chapter has focused on how and why energy is used. In particular, we have highlighted the considerable complexities involved in: (a) measuring patterns of global energy demand and their variation over space and time, and (b) reducing energy consumption levels in response to the growing pressures of climate change, security concerns and cost. Throughout, we have stressed the importance of treating the dynamics of energy demand not as largely reducible to technical or financial issues, but rather as an expression of the complex social practices within which energy consumption is embedded. Accordingly, we have finished the chapter by reflecting on the limitations of the most dominant approaches promoted as mechanisms for achieving reductions in global energy demand (for example, technical innovations, the use of incentives and the setting of national/international standards). We have stressed instead the need to place these interventions within the context of an engagement with much more fundamental questions about energy demand, such as the definition of 'need' within the context of energy service provision, and emerging attempts to articulate alternatives to hegemonic growth-oriented discourses.

QUESTIONS FOR DISCUSSION

- How would you go about defining what a basic need for energy is, and how might this vary between societal contexts around the world?

- What are the limitations of focusing energy demand reduction solely on improving energy efficiency?

- What challenges can you see in promoting a degrowth agenda?

- What is the most appropriate way of assigning carbon emissions to countries – in terms of territory, production or consumption?

- Why do you think the take-up of 'improved,' more efficient cookstoves has been relatively limited?

ACTIVITIES/POTENTIAL RESEARCH PROJECTS

- Discuss with an older person that you know how their use of and dependence on energy has changed since they were a child. What were they used to living with in terms of, say, the provision of heat, light and plugged in devices and appliances in the home? What does their account tell you about processes of social and technical change, and how they might evolve in the future?

- Consider several different definitions for a standard or 'basic' level of access to 'modern' energy services. Select one of these definitions and justify your choice in relation to the services covered. Now choose two contrasting countries from different regions of the global South and compare their current levels of basic energy access according to your selected definition. What difficulties may be encountered in trying to meet the global challenge of providing basic energy access for the whole population of both countries by 2030? Outline alternative strategies whereby these targets might be met in both countries and evaluate their likely effectiveness.

RECOMMENDED READINGS

■ Escobar, A. 2015. Degrowth, postdevelopment, and transitions: a preliminary conversation. *Sustainability Science* 10: 451–462.

This is an intriguing paper that brings together some of the themes explored in this chapter in an innovative and engaging way. Escobar attempts to initiate a dialogue between the separate literatures on degrowth and post-development, which he argues share a rejection of the core assumptions of the continuous pursuit of growth and development and a focus on the articulation of a radical, socially and ecologically sustainable alternative vision.

■ Lutzenhiser, L. 2014. Through the energy efficiency looking glass. *Energy Research and Social Science* 1: 141–151.

Energy efficiency is approached critically considering the narrow model of energy use and energy saving that is typically embedded in policy and in the work of what is characterised as the 'energy efficiency industry'. It is argued that the standard techno-economic model is problematic conceptually, as well as in terms of how it limits the scope of climate action, hampering best efforts and misdirecting policy attention.

■ Pachauri, S. 2011. Reaching an international consensus on defining modern energy access. *Current Opinion in Environmental Sustainability* 3(4): 235–240.

This paper explores the growing movement towards the adoption of a universal energy access target or goal. It does this by considering alternative definitions of access, as well as what standards for measuring access might be utilised from an operational point of view. It reflects on the mix of energy services that might be included within any definition of basic needs, the difficulties involved in setting minimum thresholds and the cost implications for families of transitioning to minimum basic energy service levels.

■ Walker, G., N. Simcock and R. Day. 2016. Necessary energy uses and a minimally-decent standard of living in the UK: energy justice or escalating expectations? *Energy Research & Social Science* 18: 129–138.

This paper examines how energy uses are included in the outcomes of deliberations by members of the public about how to define a minimally decent living standard in the UK. Energy uses are judged as necessary to enabling access to multiple valued energy services and for reasons related to health, social participation, development and practical living. While public deliberations about necessities can be taken as legitimate grounding for defining minimum standards and the scope of fuel poverty policy, they can also reveal the escalation of norms of energy dependency.

REFERENCES

Ackermann, M.A. 2002. *Cool comfort: America's romance with air conditioning.* Washington DC: Smithsonian Institution Press.

Adbusters 2017. Buy Nothing Day. Available online at www.adbusters.org/bnd/

Araújo, K. 2014. The emerging field of energy transitions: progress, challenges, and opportunities. *Energy Research and Social Science* 1: 112–121.

Aronczyk, M. 2005. 'Taking the SUV to a place it's never been before': SUV ads and the consumption of nature. *InVisible Culture: An Electronic Journal for Visual Culture* (IVC), Issue 9. Available online at www.rochester.edu/in_visible_culture/Issue_9/aronczyk.html

Barrett, J.R., G. Peters, T. Weidmann, K. Scott, M. Lenzen, K. Roelich and C. Le Quere. 2013. Consumption-based GHG emissions accounting in climate policy: a UK case study. *Climate Policy* 13 (4): 451–470.

Bhatia, M. and N. Angelou. 2015. *Beyond connections: energy access redefined.* ESMAP Technical Report; 008/15. Washington DC: World Bank. Available online at https://openknowledge.worldbank.org/handle/10986/24368

Bladh, M. 2012. Energy consumption and stocks of energy-converting artefacts. *Energy Policy* 43: 381–386.

Brand, C. and B. Boardman. 2008. Taming of the few – the unequal distribution of greenhouse gas emissions from personal travel in the UK. *Energy Policy* 36: 224–238.

Brown, E., J. Leary, G. Davies, S. Batchelor and N. Scott. forthcoming. eCook: what behavioural challenges await this potentially transformative concept? *Sustainable Energy Technologies and Assessments*, ISSN: 2213–1388.

Chitnis, M., S. Sorrell, A. Druckman, S.K. Firth and T. Jackson. 2013. Turning lights into flights: estimating direct and indirect rebound effects for UK households. *Energy Policy* 55: 234–250.

Cooper, G. 1998. *Air-conditioning America: engineers and the controlled environment, 1900-1960.* Baltimore, MD: John Hopkins University Press.

Crewe, E. 1997. The silent traditions of developing cooks. In *Discourses of development: anthropological perspectives.* R.D. Grillo and R.S. Stirrat (eds). London: Berg.

D'Alisa, G., F. Demaria and C. Cattaneo. 2013. Civil and uncivil actors for a degrowth society. *Journal of Civil Society* 9 (2): 212–224.

Davis, S. and K. Caldeira. 2010. Consumption-based accounting of CO_2 emissions. *Proceedings of the National Academy of the Sciences* 107(12): 5687–5692.

Escobar, A. 1995. *Encountering development: the making and unmaking of the third world.* Princeton, NJ: Princeton University Press.

Escobar, A. 2015. Degrowth, postdevelopment, and transitions: a preliminary conversation. *Sustainability Science* 10: 451–462.

Fouquet, R. and P.J.G. Pearson. 1998. A thousand years of energy use in the United Kingdom. *Energy Journal* 19(4): 1–41.

Friman, M. and M. Hjerpe. 2015. Agreement, significance, and understandings of historical responsibility in climate change negotiations. *Climate Policy* 15(3): 302–320.

Gent, D. 2014. *Governing energy in Nicaragua: the practices and experiences of off-grid solar energy technologies.* PhD dissertation. Loughborough University.

Hirsh, R. F. and C.F. Jones. 2014. History's contributions to energy research and policy. *Energy Research & Social Science* 1: 106–111.

Hitchings, R. 2011. Researching air-conditioning addiction and ways of puncturing practice: professional office workers and the decision to go outside. *Environment and Planning A* 43(12): 2838–2856.

Hitchings, R. and S.J. Lee. 2008. Air conditioning and the material culture of routine human encasement: the case of young people in contemporary Singapore. *Journal of Material Culture* 13(3): 251–265.

Jeroen, C. J., M. van den Bergh, and G. Kallis. 2012. Growth, a-growth or degrowth to stay within planetary boundaries?. *Journal of Economic Issues* 46(4): 909–920.

Kallis, G. and H. March. 2015. Imaginaries of hope: the utopianism of degrowth. *Annals of the Association of American Geographers* 105(2): 360–368.

Lovins, A.B. 1990. The negawatt revolution, *Across the Board* XXVII(9): 21–22.

Lutzenhiser, L. 2014. Through the energy efficiency looking glass. *Energy Research and Social Science* 1: 141–151.

Martinez-Alier, J., U. Pascual, F.D. Vivien and E. Zaccai. 2010. Sustainable de-growth: mapping the context, criticisms and future prospects of an emergent paradigm. *Ecological Economics* 69(9): 1741–1747.

Mortimer, K., C.B. Ndamala, A.W. Naunje, J. Malava, C. Katundu, W. Weston, D. Havens, D. Pope, N.G. Bruce, M. Nyirenda and D. Wang. 2017. A cleaner burning biomass-fuelled cookstove intervention to prevent pneumonia in children under 5 years old in rural Malawi (the cooking and pneumonia study): a cluster randomised controlled trial. *The Lancet* 389(10065): 167–175.

Moss, T. 2015. SDG seven: update the 'modern' in universal modern energy access. Available online at www.cgdev.org/blog/sdg-goal-seven-update-modern-universal-modern-energy-access

Neven, L., G. Walker and S. Brown. 2015. Sustainable thermal technologies and care homes: productive alignment or risky investment? *Energy Policy* 84: 195–203.

Pachauri, S. 2011. Reaching an international consensus on defining modern energy access. *Current Opinion in Environmental Sustainability* 3(4): 235–240.

Preston, I., V. White, J. Thumim and T. Bridgeman. 2013. *Distribution of carbon emissions in the UK: implications for domestic energy policy.* York: Joseph Rowntree Foundation.

Rafey, W. and B.K. Sovacool. 2011. Competing discourses of energy development: the implications of the Medupi coal-fired power plant in South Africa. *Global Environmental Change* 21(3): 1141–1151.

Revkin, A. 2011. The ad man behind Occupy Wall Street and 'Buy Nothing Day'. *New York Times*, 25 November 2011.

Rollins, W. 2006. Reflections on a spare tire: SUVs and postmodern environmental consciousness. *Environmental History* 11 (October 2006): 684–723.

Røpke, I., H.C. Toke and J.O. Jensen. 2010. Information and communication technologies – a new round of household electrification. *Energy Policy* 38(4): 1764–1773.

Sahakian, M. 2011. Understanding household energy consumption patterns: 'When West is best' in Metro Manila. *Energy Policy* 39(2): 596–602.

Sayer, A. 2011. *Why Things matter to people: social science, values and ethical life.* Cambridge: Cambridge University Press.

Scott, A. and C. Miller. 2016. Accelerating access to electricity in Africa with off-grid solar: the market for solar household solutions. Overseas Development Institute. Available online at www.odi.org/publications/10200-accelerating-access-electricity-africa-off-grid-solar

Shove, E. 1998. Gaps, barriers and conceptual chasms: theories of technology transfer and energy in buildings. *Energy Policy* 26, 1105–1110.

Shove, E. and G. Walker. 2014. What is energy for?: energy demand and social practice Theory, *Culture and Society* 31(5): 41–58.

Shove, E., Walker, G. and Brown, S. 2014. Transnational transitions: the diffusion and integration of mechanical cooling. *Urban Studies* 51(7): 1504–1517.

Sivak, M. 2009. Potential energy demand for cooling in the 50 largest metropolitan areas of the world: implications for developing countries. *Energy Policy* 37: 1382–1384.

Sørensen, B. 2012. *A history of energy: Northern Europe from the Stone Age to the present day.* Abingdon: Earthscan.

Torriti, J. 2016. *Peak energy demand and demand side response.* London: Routledge.

Urry, J., 2010. Consuming the planet to excess. *Theory, Culture & Society,* 27(2–3): 191–212.

Urry, J. 2013. *Societies beyond oil: oil dregs and social futures.* London: Zed Books.

Walker, G. 2014. Dynamics of energy demand: change, rhythm and synchronicity. *Energy Research and the Social Sciences* 1: 49–55.

Walker, G. 2016. *De-energising and de-carbonising society: making energy (only) do work where it is really needed.* Friends of the Earth Big Ideas Series. Available online at www.foe.co.uk/sites/default/files/downloads/de-energising-society-102383.pdf

Walker, G., E. Shove, and S. Brown. 2014. How does air conditioning become 'needed'? A case study of routes, rationales and dynamics. *Energy Research and Social Science* 4, 1–9.

Walker, G., A. Karvonen and S. Guy. 2015a. Zero carbon homes and zero carbon living: sociomaterial interdependencies in carbon governance. *Transactions of the Institute of British Geographers* 40(4): 494–506.

Walker, G., A. Karvonen and S. Guy. 2015b. Reflections on a policy denouement: the politics of mainstreaming zero-carbon housing. *Transactions of the Institute of British Geographers* 41(1): 104–106.

Walker, G., S. Brown and L. Neven. 2015c. Thermal comfort in care homes: vulnerability, responsibility and 'thermal care'. *Building Research and Information* 44(2): 135–146.

Walker, G., N. Simcock and R. Day, 2016. Necessary energy uses and a minimally-decent standard of living in the UK: energy justice or escalating expectations?. *Energy Research & Social Science* 18: 129–138.

White, L.A. 1943. Energy and the evolution of culture. *American Anthropologist* 45: 335–356.

Wiig, A. 2017. Charging smartphone batteries, powering the internet: conceptualising the conjoined energy infrastructures of mobile, digital connectivity. In *Demanding energy: space, time and change*. A. Hui, R. Day and G. Walker (eds), London: Palgrave Macmillan.

Wilhite, H. 2008. New thinking on the agentive relationship between end-use technologies and energy-using practices. *Energy Efficiency* 1: 121–130.

World Economic Forum. 2015. *The global information technology report 2015*. Geneva: World Economic Forum.

Energy controversies and conflicts

Controversies and conflicts are endemic to energy landscapes around the world. From the slow submerging of low-lying islands due to sea-level rise induced by climate change, through to the sudden experience of major energy-related disasters, the uneven distributions of well-being and harm associated with energy landscapes are clear. So, too, is their contested and intrinsically political character. Conflicts can erupt, sometimes with intense ferocity, around specific plans for implementing energy capture and conversion technologies at particular sites, such as shale gas drilling, wind farms, gas pipelines or nuclear power stations. They can also centre on future planning processes for national energy systems and the technologies or governance models they strategically favour, or proposed costs and speed of energy system transformation. Conflicts may focus, too, on the direct and localised consequences of energy generation, distribution or consumption, such as from mining operations, oil spills or urban traffic pollution; or on the spatially and/or temporally distant consequences of carbon-induced climate change, and the need to mitigate potentially catastrophic impacts through rapid energy system transformation. Controversy and conflict are seemingly everywhere in energy landscapes, and not only as part of how they have developed historically: political tensions and contradictions continue to be central to how future energy landscapes are being imagined and shaped today.

A critical social science approach draws attention not just to the topics of contestation, but also to their social unevenness: there are always winners and losers in energy landscapes, although their distribution can change over time. Typically, the problematic consequences of energy systems play into existing patterns of social and environmental inequality and injustice and reflect wider contours of power and agency. Conflicts are often characterised by distinct asymmetries in the power, influence and resources available to different actors. Powerful state and industry actors typically work to develop, sustain and reproduce particular forms of energy systems which, given the centrality of energy to wider processes of neoliberal capitalism, means disruptions to energy systems or to planning processes for their development are rarely tolerated. Communities and/or activist movements that seek to challenge incumbent interests and offer energy alternatives can, as a consequence, find it difficult to make headway as they frequently have fewer resources to be heard and less capacity to influence.

The prevalence and intensity of contestation over energy issues reflects, in part, the outcome of scientific investigation of energy-related patterns of pollution, risk and other forms of impact on human health and environmental change, and the uncertainties and complexities this investigation often produces. But controversy rests, too, on different ideas about fairness and justice, for now and in the future, and on what constitutes a sufficiently secure and sustainable energy system. In this chapter, therefore, we explain how energy controversies involve forms of claim-making that extend across matters of evidence and knowledge, and include claims that have a more overtly ethical, normative and political character. Controversy and conflict, therefore, are not just problems to be avoided: they are inherent to deliberation and the democratic political process, and can have a productive role in questioning how energy systems should be developed and ensuring proper scrutiny in the decision-making process.

The chapter is divided into three sections. The first considers the main topics of controversy that have emerged across different parts of energy systems, covering health, environmental, economic and landscape impacts, and questions of energy and carbon performance. Examples demonstrate the multiplicity of concerns across different spatial and temporal scales, and how these can combine together and emerge in particular physical, political and cultural settings. In the second, the notion of claim-making is introduced and the varied forms of claim-making that run through energy controversies are examined, from the 'boundary work' of scientific knowledge claims, through to ethical and political claims for rights of participation and recognition. The final section reflects on the underlying processes that generate forms of social conflict around energy questions, considering competing explanations (NIMBYism through to critiques of neo-liberal growth models) and the alternative forms of response and prognosis they suggest. The move towards using 'volunteering' mechanisms in energy facility siting processes is used to exemplify some of the tensions involved, with links made through to the transition processes discussed in Part 3 of the book. Across the chapter a diversity of forms and cases of controversy and conflict are considered. There are also many other instances where energy landscapes are the subject of dispute and contestation in this book, so learning from this chapter can be transferred to analyse other contexts and cases.

CHARACTERISING ENERGY CONTROVERSIES

An energy controversy has been defined by Boucher (2012: 149) as "a situation in which actors, participants . . . engage in politically salient discursive conflict". They are the places and times when one or more of the socio-political tensions inherent to energy landscapes becomes openly expressed. These 'situations' may be spatially focused and bounded, or extend across and between many different locations and settings. They can also be relatively short-lived, with the conflict focused and resolved (or at least ended) relatively rapidly; or long-lasting and unresolved for extended periods of time, maybe following an episodic ebb and flow of visibility and politically saliency.

Walker et al. (2011) engage with some of these temporal dynamics in their general framework for characterising interactions between local publics and project developers – specifically in relation to renewable energy technologies, such as wind farms and **energy from biomass** plants, in the cases they researched. Their framework (see Figure 7.1) is based around evolving iterative loops of expectations being formed, engagement actions being undertaken (such as public meetings or protest marches) and expectations being reshaped as a consequence. These interactions build over time, punctuated by key events and decision stages. They note that even processes leading towards specific and spatially focused 'yes or no' siting decisions can take many years to be resolved for controversial projects: for example, it took five years in total for a proposed energy from biomass development they tracked over time. Such timescales are in tension with demands to materialise low-carbon energy sources at scale and at speed.

Controversy and conflict are often cast in negative terms, because they can complicate and delay energy system planning and implementation, and consensus and agreement are frequently the preferred position. However, this logic is challenged by political theorists who argue that democratic systems are ones in which resolute differences of perspective are to be expected and actively worked with. For example, the political theorist Chantal Mouffe (1998) argues that "while we desire an end to conflict, if we want people to be free we must always allow for the possibility that conflict may appear and to provide an arena

energy from biomass
biomass is a general term for organic matter such as plants and trees as well as waste from food and agricultural systems. There are a variety of ways of extracting energy from biomass: generating electricity: through forms of incineration, through conversion into a bio-fuel that can be used as a petrol or diesel substitute: or through conversion into biogas. Extracting energy from biomass has been promoted as both sustainable and low carbon, but these designations have been subject to challenge and controversy.

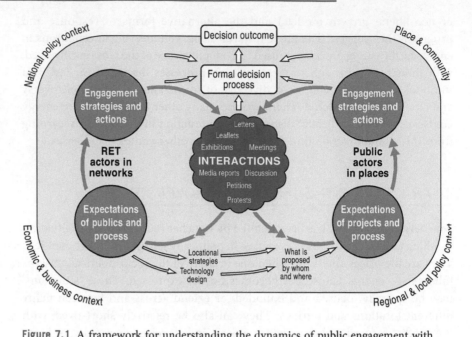

Figure 7.1 A framework for understanding the dynamics of public engagement with renewable energy projects (Walker et al. 2011)

where differences can be confronted. The democratic process should supply that arena". In such instances we might well expect controversies to be longer lived, rather than readily resolved.

An example of an ongoing controversy that has extended widely over time and space are the intense debates over the merits of nuclear power. These have circulated ever since nuclear energy's first use for civil purposes in the 1950s, although the form and pattern of controversy have shifted substantially over time, particular in response to accidents and releases of radioactivity, for example at Chernobyl in 1986 (Williams and Baverstock 2006) and Fukushima in 2011 (Wittneben 2012). The nuclear debate has also been strongly differentiated over space. As a result of strong public opposition, some countries such as Germany have now set themselves against nuclear energy and are actively

CASE STUDY: IMPACTS AND SOCIAL CONSEQUENCES OF THE DEEPWATER HORIZON OIL SPILL

In April 2010, an oil rig conducting exploration drilling around 40 miles (60 km) offshore in the US part of the Gulf of Mexico suffered a major blowout. An explosion rendered the pressure control systems on the Deepwater Horizon rig redundant, leading to a fireball that engulfed the platform and killing eleven workers. The oil rig was owned by Transocean, one of the world's largest offshore drilling contractors, and was contracted to drill the Macondo oil prospect managed by BP, the London-based transnational oil company. The resulting fire could not be immediately extinguished, and there was an uncontrolled release of an estimated 4.9 million barrels of oil into the Gulf of Mexico over 89 days, the biggest environmental disaster in US history and the world's largest marine spill. The impacts and consequences of the Deepwater Horizon disaster have been far reaching and the subject of much controversy.

In the period immediately following the disaster, there were divergent assessments of exactly how much oil was being released and significant uncertainties associated with all the numbers generated. There were also different assessments of how far oil was being distributed within the marine environment, and where its impacts were being felt. In the early days of the crisis, a non-profit organisation that uses satellite images to gather evidence about the environmental consequences of human activity across the planet (SkyTruth), successfully challenged BP and federal official estimates of the volume of oil being released. Their challenge resulted in a revised estimate some five times higher (Cart 2010). Many local people also distrusted official modelled assessments of where oil was coming ashore and impacting wildlife, to the degree that alternative citizen-led processes of evidence gathering were deployed. Coordinated by the Public Laboratory for Open Technology and Science (Breen et al. 2015), local groups used cameras rigged to kites and balloons to produce extensive photographic maps of spill-impacted areas. Other projects accumulated data collected by local people on instances of birds contaminated with oil. The alternative body of evidence amassed by these groups told a different story about the extent of environmental damage, and contributed to public debate over the disaster's ecological and economic consequences.

The blowout's economic consequences quickly rippled through coastal fishing communities affected by the release of oil, and associated trades and industries. Calculating the costs to businesses and residents of lost income arising from the spill has been a major source of disagreement and controversy, and has led to protracted court battles. Many of the coastal communities in Louisiana, Mississippi and Alabama that suffered lost income because of fishing restrictions following the disaster were already economically disadvantaged, so that the effects of Deepwater Horizon fell disproportionately on those with limited economic alternatives (Safford et al. 2012). The Deepwater Horizon oil spill also raised important questions about gaps in the regulation of oil exploitation activity and the enforcement of existing standards; about how the pursuit of profit can override health, safety and environmental concerns; and about the consequences of pursuing 'extreme energy' (**Chapter 1**). Critics pointed to the ways in which corporate social responsibility is framed, and often disregarded, in parts of the energy industry. BP were widely criticised for 'greenwashing', promoting an image of environmental stewardship around a shift to renewables while actively expanding their fossil reserves, and apparently cutting corners to meet production goals and costs (Matejek and Gossling 2014). In 2015 BP agreed to pay $18.7 billion in fines, the largest corporate settlement in US history. The company's total bill, including multiple settlements with federal and state authorities, property owners and shareholders, is over $60 billion. The fact that BP is able to deduct some of these costs as a normal business expense when calculating their annual tax liability – meaning the US taxpayer effectively subsidises the company's clean-up bill – has been another source of controversy.

seeking to phase it out (see **Chapter 9**); while other countries, such as France and the UK, are seeking to incorporate nuclear power within a 'low carbon' energy profile (Pidgeon et al. 2008). Many local-scale and shorter-lived disputes have erupted and progressed within this wider flow and flux of nuclear controversy, amounting in total to a complex and variegated picture of ongoing and enduring contestation.

There are potentially very many different sources of energy conflict. Table 7.1 lists a set of generic categories that can provide the focus of 'discursive conflict'. This list identifies some of the main and recurrent topics of contestation, although it is not intended to be exhaustive. We discuss these topics separately below, but it is important to recognise that they are usually argued over in combination and/or relation to each other. Where and how they arise is also contingent on the particular focus of contestation (e.g. the technologies and places involved); the profile of the actors who become involved in discursive exchanges; the scale at which the dispute emerges and over which it subsequently comes to circulate; and the wider political or economic context in which they are set. **Example 1** (on energy developments in marine spaces), **Example 2** (on hydropower and conflict in Chile) and a **Case study** on the Deepwater Horizon oil spill exemplify these ideas.

Health consequences are perhaps the most charged focus of energy controversy, given the importance of health to fundamental dimensions of human well-being. They can arise across all the different component parts of energy systems and extend across space and time. As a result, energy systems can have health implications for many different population groups. Examples include:

environmental justice
the notion of environmental justice first emerged in the US in the 1980s, around protest against the unequal socio-spatial distribution of waste and toxic sites. It has subsequently evolved and developed as a political frame in many other places around the world. For Walker (2012) it is about the 'entwining of environment and social difference' that can include many different energy-related concerns – pollution, climate change, mineral extraction, and fuel poverty for example – and many different forms of social difference, including deprivation, race, ethnicity, gender and age.

- the risks to workers involved in the extraction of primary energy resources (coal, oil, gas and uranium);
- the health impacts of pollution emissions from power stations both locally and at regional scales;
- the health risks distributed along transmission and distribution systems e.g. from overhead power lines or major gas pipelines;
- the health consequences of people in poverty cooking with low grade fuels;
- the excessive winter deaths arising from people being unable to heat their homes to an adequate level (see **Chapter 5**).

Health is also a major concern in terms of climate impacts (Levy and Patz 2015). In many cases it is not just a specific health impact or risk that is at issue, but their accumulation in particular places and/or for people who are especially vulnerable to harm. Increasingly, these cumulative and negative effects of energy systems are being interpreted as a matter of **environmental injustice** (Walker 2012; Bickerstaff et al. 2013). This framing highlights the disproportionate and unfair health burdens experienced by people in particular localities (e.g. with a concentration of energy infrastructures); or, more broadly, by people

Table 7.1 Examples of topics of controversy and conflict related to energy systems

Category	Explanation	Examples
Health consequences	Impacts on human health, both physical and psychological typically from forms of pollution, risk or contamination	Pollution from fossil fuel or nuclear power stations; risks of accidents within gas production and distribution systems; indoor pollution from burning dirty fuels; impacts on workers in resource extraction
Environmental and ecological consequences	Impacts on flora and fauna, ecosystems and biodiversity and on atmospheric systems	Pollution from energy production; landscape destruction from mining; impacts of oil spills in marine environments; changes to tidal environments from barrages
Economic consequences	Impacts on economic growth, jobs and employment and on energy prices	New jobs generated from investments in oil production or fracking; consequences for tourism and house prices from wind farms; effects of low carbon technology subsidies on energy prices
Landscape and visual impacts	Impacts on landscape quality and on senses of place associated with valued landscapes	Landscape and visual impacts of wind farms and electricity pylons; landscape consequences of open cast mining and hydroelectric dams
Civil liberties and human rights	Consequences for rights of freedom, political participation and self determination	Imprisonment and violence against protestors; secrecy associated with the nuclear industry; ownership and security of data from smart meters
Energy performance	The amount and reliability of energy generated, the comparative costs of production, and patterns of loss or waste of energy	The intermittency of solar or wind generation; the loss of energy in conversion from gas to electricity; the full costs of nuclear power including long term waste disposal
Carbon performance	The carbon production associated with different energy systems and policies	The effectiveness of carbon capture and storage; the carbon footprint of the full nuclear cycle; the carbon balance of biofuel feedstocks

that are poor, marginalised politically or from racially or ethnically defined communities.

Environmental and ecological consequences are sometimes closely tied to health concerns, but are also a distinct category of controversy where impacts on flora and fauna, on ecosystems and biodiversity, and on the functioning of wider natural systems are at issue. The full range of an energy system's components can have environmental and ecological consequences, from resource extraction to waste disposal. These extend from very localised impacts on particular ecological spaces and species; to region-wide and transnational consequences, such as impacts at the watershed scale or the extended transmission line impacts of large-scale hydroelectric schemes (Schaeffer and Smits

2015); and to global scale ecological shifts, disruptions and damage, such as those already being experienced and/or predicted as a consequence of anthropogenic climate change strongly associated with the combustion of fossil fuels. Questions about what should be valued, for whom and why are often central to these conflicts, setting the eco-centric concerns of environmentalists against those with a distinctly **anthropocentric** focus on the contribution of energy systems to livelihoods and economic development.

Economic consequences of energy systems often become a focal point of controversy in both positive and negative ways, and are frequently contested in those terms. Proponents of unconventional oil and gas development, for example, such as fracking for oil and gas (see Figure 7.2), deploy arguments about job creation, tax income (locally and nationally) and reductions in energy prices (Cotton 2016). In Europe, positive economic experiences from the US fracking industry have been used to back up such claims; while opponents counter by arguing that the US case is very particular, that it cannot be readily replicated in Europe where both geology and land tenure are significantly different, and that economic expectations are wildly overblown (Bomberg 2017a).

anthropocentric
in a literal sense this means 'human-centred', and is used to convey a way of seeing the world that foregrounds the well-being and interests of humans and society. It is often contrasted with an 'ecocentric' world view, which either gives equal priority to non-human living entities – such as animals, plants and ecosystems – or in some case prioritises them over human interests.

EXAMPLE 1: ENERGY DEVELOPMENTS IN MARINE SPACES

Marine environments are becoming increasingly important spaces for new energy investment and the development of a range of energy infrastructures. Roberts (2013) examines three examples of proposed energy developments in marine settings around the UK coast that are the subject of controversy – an offshore wind farm, a wave power testing facility and underground CO_2 storage. Each of these gathered together varied cast of local, regional and national actors, including developers, regulators, fishermen, surfers, chambers of commerce and local politicians. Concern about environmental impacts has focused in particular on commercially valued marine ecosystems and the industries that rely upon them. In the case of an offshore wind farm, for example, concern focused on disruption to the sea bed and established fishing grounds as well as visual impacts to the seascape. For the wave power facility, concerns centred on impacts on surfing from changes to wave patterns, access to fishing grounds and safety for shipping. The main concerns in the case of the carbon storage proposal were about the uncertain risks and consequences of carbon leakages and, more generally, about dumping a form of 'waste' in a wilderness environment that should be protected for future generations. As Roberts notes, just because developments are in offshore marine spaces does not "negate the need to consider stakeholder/public acceptance" (ibid.: 126). His work found strong patterns of identification with local sea spaces that are highly valued, even where these spaces are some distance away from where people live. The value people place on sea spaces often reflects the way marine environments have been part of their working and recreational lives for many years.

Figure 7.2 Anti-fracking protestors, London (photo credit: Gary Knight)

The prospect of positive economic impacts is also often set against the negative economic consequences that can arise from losses to existing employment and income (see **Example 1**). Concern for the livelihoods of those involved in fishing and tourist industries frequently feature in debates over offshore energy development; and, given the frequently rural setting of land-based energy infrastructures, potential impacts to agriculture and tourism can be particularly contentious.

Landscape and visual impacts also relate to the particular spaces in which forms of energy infrastructure are located. These impacts are generally localised and are frequently experienced as intrusions of technology and infrastructure into areas that valued in terms of their landscape quality (or seascape quality, see **Example 1**) and sense of place. Onshore wind farms, for example, need to be located in areas with high wind speeds, which are often open and rural in character, such as coastal or hilly areas, and these locational factors explain some of the deep contestation that has emerged over their siting (Van der Horst and Toke 2009; Devine-Wright 2009). Similarly, open pit coal mining, and extractive techniques like mountain-top removal, are radical intrusions into a pre-existing landscape that transform surface features and create entirely reworked landscape forms. Large-scale hydropower projects have ignited controversy in Iceland and in Chilean Patagonia for similar reasons (**Example 2**) as they flood upland valley landscapes, and require extensive connecting infrastructure to supply electricity to distant consumers (Figure 7.3).

Figure 7.3 A poster from the 'Patagonia without Dams' campaign

Civil liberties and human rights feature in a more limited range of energy controversies and conflicts. They can, however, be particularly contentious concerns that draw in a range of non-governmental organisations and other actors nationally and internationally. The experience of oil exploitation in the Delta Region of Nigeria shows how measures put in place to secure and protect energy investments can deeply compromise the freedom and civil liberties of community organisations and other concerned groups (Omerje 2005; Watts

2004). There has been appalling direct violence in the Delta, as well as what Nixon (2011) characterises as the **slow violence** of incremental destruction of the landscapes, cultures and ways of living. Energy development projects damaging the land and livelihoods of indigenous peoples have repeatedly been cast as violations of human rights and/or the violent use of state or corporate power. Examples include the North Dakota pipeline in the US which, Native American protestors argue, will transgress sacred native land and threaten the purity of their water resources (Levin 2016); nuclear waste dumps proposed for the territory of First Nation peoples in Canada (Stanley 2008); and the submerging of several small-island states as a consequence of sea-level rise induced by climate change (Barnett and Campbell 2009). Here, nothing less than the existence of histories, cultures and identities is under threat as a consequence of the failure to decarbonise energy systems.

Shifting to the demand side of energy systems, questions of civil liberties have recently become contentious in relation to the introduction of **smart meters** into people's homes and the flows of information expected to accompanying the operation of smart grids (see **Chapter 10**). Who owns household energy data and has the right make use of it, and on what basis control might be exercised remotely over domestic appliances as part of demand response measures, have become much debated questions (Naus et al. 2014).

Energy performance may appear a relatively technical matter, but contestation, including between different expert voices, has centred on questions of cost, reliability and security of energy supplies and also on the amount of energy lost or wasted by different infrastructural systems. Much debate has centred on the **intermittency** of various forms of renewable energy (see the discussion later in this chapter), with critics arguing that wind turbines or solar panels cannot be relied on to supply energy when it is most needed. Others argue, however, that across a dispersed grid some balancing out of intermittency takes place (if the wind is not blowing in one place it will be somewhere else), and that every form of energy supply needs to be considered as part of a mix of alternative energy sources.

The *carbon performance* of particular technologies and/or policies is a relatively recent focus of energy debates. Carbon is increasingly a key metric of how energy systems are evaluated, so being able to frame a technology or policy as low or high carbon can have significant consequences. As such, carbon performance has become a subject of contestation and challenge. Proponents of nuclear power have made much headway over the last decade in repositioning it as a low or even zero carbon form of electricity generation (see discussion later in this chapter). Opponents challenge the low-carbon credentials of nuclear power, pointing to the greenhouse gas emissions involved in mining, transporting and processing uranium fuel, building power stations and

slow violence
this term is used by Nixon (2011) to capture how violence can be enacted slowly, gradually and out of sight and in a way that is dispersed over time and space. This contrasts with the sudden and immediate way violence is usually understood to unfold. Fossil-fuel based energy systems, for example, enact forms of slow violence through the effects of land degradation, air and water pollution on places and communities, as well as the long term consequences of climate change.

smart meters
digital meters that are able to record electricity flow with more precision than traditional analogue meters and transmit these readings over digital and usually wireless systems. They can also, to some degree, be managed and controlled remotely. Many countries are going through programmes of installing smart meters in households and businesses in order to then enable the introduction of time-of-use pricing and intelligent real-time balancing of supply and demand within smart grids.

intermittency
some forms of renewable energy generation operate on an intermittent basis subject to the strength of, for example, wind flow, solar radiation or wave movement. This is seen by some as a problem in contrast to the constancy of coal or nuclear powered electricity generation. Such intermittency can though now be fairly well predicted and managed within an integrated and smart energy system, but there is much disagreement about how much this can be relied on.

handling wastes (Sovacool 2008). Biomass energy – in the form of biofuels – provides another example of where a lower-carbon fuel has been at the centre of controversy. In the early 2000s biofuels were promoted, initially with little opposition, as a "silver bullet solution to concerns about the environment, energy security and rural development" (Boucher 2012 with their globally important carbon performance particularly stressed. Biofuels can be carbon neutral *if* there is balance between the CO_2 emitted during combustion and that taken from the atmosphere while the feedstocks are being grown. The presumption of carbon balance has increasingly been challenged, however, with critics pointing to the 'carbon debt' produced by clearing land for biofuel production and the knock-on impacts on locally experienced food security and soil and water degradation problems (Palmer 2014), including in areas such as Amazonia (Walker 2011).

CLAIM MAKING AND CONFLICT: SCIENCE, JUSTICE AND POLITICS

Having worked through some of the topics over which sometimes very intense contestation can emerge, we can see why energy systems are frequently a focus of conflict. Across the diversity of controversies, however, a number of recurrent forms of 'claim-making' are at work. In this section we analyse in more detail the types of claims made, and how these are shaped by strategies of the actors involved. The definition of controversy provided earlier makes it clear that controversies manifest themselves through discourse – the production, flow and circulation of words, texts and images. These discourses are the means through which claims are made about the scale and significance of the types of impacts and consequences discussed earlier, but also about the reliability of evidence being put forward by other competing voices and the grounds on which decisions should be made. Claims, then, are about competing interpretations of evidence and knowledge (of how things are), but also about normative questions (how things ought to be) and about the interplay between them (Walker 2012).

In the following discussion, we show how making claims involves various strategic choices. Claimants can try to get 'science' on their side, laying claim to the 'truths' and reliability generally associated with scientific knowledge. Gieryn (1983, 1999) refers to this as 'boundary work' in which actors attempt to draw demarcations between science and non-science, rationality and irrationality, or sometimes between good science and bad science. Involved actors can also make strategic choices in terms of the scale(s) at which claims are constructed in order to gain support. Parties to a controversy can use 'scale shifting' as a tactic to make seemingly local matters into national or international ones, or sometimes vice versa (Kurtz 2002). There are also strategic choices made about how to view the processes of decision-making that are available: whether to respect these processes, and engage in the opportunities that are available for different stakeholders to be involved; or to critique the procedures being followed, and challenge the legitimacy of decisions taken – maybe even withdrawing from participating at all. Regardless of the tactics chosen, the way knowledge claims circulate is often heavily mediated by conventional news networks and social media, which selectively take up and magnify discourses in particular ways. The examples that follow exemplify these points.

Climate change is a good illustration of how controversies over scientific knowledge claims at the global scale can be used to advance arguments about the direction of energy policy at national and local scales. Research by authors such as Barr (2011), Stoll-Kleemann et al. (2001) and Whitmarsh (2009) illustrates how several different discourses exist around climate change and climate science. These have been linked to the continuing scepticism and denial of climate change expressed by some policymakers, politicians and political affiliates, particularly after the disclosure of data and correspondence from the University of East Anglia's Climate Research Unit in late 2009. These appeared to show that climate data were being manipulated, an accusation subsequently found to be unfounded after an inquiry. Known as 'Climategate' (Leiserowitz et al. 2013), this episode was for a while hugely damaging to the public perception of peer-reviewed climate science and provided new grounds for those who oppose measures to mitigate climate change, despite the clear findings of the most recent IPCC (2014) report on the impact on the climate of human activity. Climategate illustrates the process of scale-shifting, in that a set of internal email conversations within the scientific community became a globalised knowledge controversy.

Knowledge claims around climate change are based on peer-reviewed science about the impact of human activity on the climate, and on the need for rapid and radical change in policies to tackle resultant warming. Research by Barr (2011) highlights how public scepticism over the practices of climate researchers has led to an overall decline in public trust in the scientific process (Leiserowitz et al. 2013). However, ideology plays a role too, as can be seen by the ways in which claims about climate change are mobilised in the popular media. Carvalho's (2007: 223) analysis of climate change reporting in British 'quality' newspapers found that:

> Understood . . . as a set of ideas and values that legitimate a program of action vis-à-vis a given social and political order, ideology works as a powerful selection device in deciding what is scientific news, i.e. what the relevant 'facts' are, and who are the authorized 'agents of definition' of science matters.

The point here is that, in conveying particular claims, conventional and social media (re)interpret the work of scientists. In doing so, they re-draw the boundaries between what is considered legitimate and problematic forms of scientific practice (Gieryn 1999), and the implications of research findings for economic and political decision-making. Accordingly, Carvalho (2007: 223) argues that the "representation of scientific knowledge has important implications for evaluating political programs and assessing the responsibility of both governments and the public in addressing climate change". Public controversies over climate change reflect the way different actors make claims about the science that is used to create and implement policy, and the ways in which scientific knowledge is interpreted in wider society. Claims about scientific knowledge can readily be mobilised to either support or to argue against particular energy policies. For example, the right-of-centre *Daily Mail* newspaper in the UK has ably mobilised its sceptical position on the issue of climate change to strongly argue against renewable energy developments, on the grounds that they are unnecessary and represent a 'green tax' on British citizens based on 'bad' science (*Mail on Sunday*, 17 March 2013).

Many knowledge controversies may have global dimensions but are typically worked out in specific localities, and often in relation to the siting or expansion of energy facilities. In recent years, the siting of shale gas drilling, wind turbines, solar farms, hydroelectric plants and tidal barrages has sparked controversy and conflict in a wide range of local settings across the globe. Taken together, these instances of dispute highlight the diverse character of knowledge claims associated with energy siting controversies, and how they frequently revolve around competing claims about the use of scientific data, issues of legitimacy and governance, and the values associated with particular landscapes. In a study of public beliefs about wind energy, Devine-Wright et al. (2006: 243) found that many different groups mobilise scientific data to support their arguments but that, in doing so, each group also undertakes various types of 'boundary work'. For example, both supporters and opponents of wind energy developments focused on the issue of energy supply reliability, although

> sceptics made intermittency familiar using concepts of unpredictability and uncontrollability, as well as notions of work over idleness; while supporters referred to variability and fluctuation. Specific themata suggest, despite polarised views, a common underlying structure to social representations of intermittency, centring on compatibility with the grid, views of the weather, the distinctiveness of wind energy and the virtue of facts over myths.

In this way, both groups make use of research data that suggested energy supply from wind turbines would be intermittent, but this was discursively represented in different ways.

In addition to competing knowledge claims, controversies can arise from a lack of meaningful engagement with affected communities (Lesbriel and Shaw 2000). Wind energy projects, for example, are typically either sited by using legislation that assigns land for such developments, or by the application of simple market principles to site development and identifying areas likely to yield the highest value return on investment. These approaches can lead to a stand-off between the site developer and local community groups, in which an apparent binary choice – for versus against – oversimplifies debate and renders consultation processes inadequate. Indeed, poor governance of wind energy projects has led developers to privilege some community organisations and forms of political engagement over others, in ways that can completely de-legitimise the process for developing new renewable energy capacity. It is frequently the case that processes of knowledge acquisition and knowledge sharing are unequal and so, to many, they are regarded as unjust. As a consequence, wind turbines are often understood by local communities as being 'imposed' from outside.

Value conflicts associated with renewable energy developments often revolve around oppositional binaries – e.g. natural environment vs industrial construc-tion – as different stakeholders debate environmental values alongside social and economic imperatives. However, issues are frequently more complicated than such discursive binaries allow, as illustrated by Jackson's (2011) study of a hydroelectric dam in Portugal and a tidal barrage in the UK. These two cases illustrate how conflict can also emerge from differences between nature conservation values at the local level (such as concerns over biodiversity and habitat loss) and international imperatives associated with reducing carbon emissions through deploying large-scale renewable energy schemes. Jackson's

---

EXAMPLE 2: HYDROPOWER CONFLICTS IN PATAGONIA, CHILE

The energy system in Chile is institutionally very centralised, dominated by big companies that monopolise generation and distribution, and with strong central state involvement in major infrastructure projects. In 2005 people in Aysén, Patagonia, became aware of plans to build new hydro-electric power (HEP) facilities consisting of five large dams in the region. A coalition of organisations formed the Patagonia Defense Council (Consejo de Defensa de la Patagonia) and began immediately to actively oppose these plans. A broad social movement grew in opposition to the dams and sustained a campaign – 'Patagonia sin Represas/Patagonia without Dams' – until 2014. Protest activity was intense and locally focused, but drew on a wide network of community groups and non-governmental organisations that extended nationally and internationally, including groups like International Rivers, Natural Resources Defense Council and Greenpeace (Schaeffer and Smits 2015). Arguments against the major development plans focused not only on the direct destruction of highly valued landscapes and native species, but also the 2,300 km long transmission infrastructure to be built to connect up to the regional grid system in the north of Chile and big mining companies that were to be the major consumers (see Figure 7.3). The projects were, therefore, seen as primarily in the interests of corporate and national state actors, rather than for the benefit of local communities. The opposition was ultimately successful with the government denying permission for the HEP projects on environmental grounds in 2014. Local action grew out of this conflict focused on enacting an alternative, sustainable and green energy pathway that embodied local decision-making and participation and independence from the central state (Walker and Baigorrotegui 2016).

(2011) study, along with many others, highlights the ways in which conflicts over energy development frequently combine knowledge controversies, debates over processes of public participation, and fundamental question of value (see **Example 2**).

Beyond site-specific conflicts, the way energy controversies are structured can also be seen at the national and supra-national scale. Bomberg's (2017b: 115) analysis of attempts to promote fracking for shale gas in the United States and the European Union provides an illustration of how the political 'framing' of energy issues can make a fundamental difference to the outcome of applications to frack for gas:

> US proponents of fracking have been more successful in delivering the reassurance frame to help maintain a comparatively lax regulatory framework. Not only did they prioritise this message, but underlined its credibility by repeated reference to the US' rich experience with drilling and innovative technology. Conversely, with their heavy emphasis on fossil fuel lock-in, the European anti-fracking network successfully exploited European citizen's greater concern with climate change, and the EU's institutional desire to play a global climate role ... The US pro-fracking network has been initially more successful than its European counterpart because it invoked a powerful idea (visible economic gain) at the right time (early on, before environmental concerns mounted).

Accordingly, the way energy development is framed by reference to national or supra-national discourses, and the way it mobilises the network of actors associated with different positions, can have a fundamental impact on the outcome. This demonstrates the important influence of underpinning discourses and sociotechnical imaginaries (see **Chapter 4**) on how energy controversies may be out-worked locally.

The examples of claim-making discussed in this section illustrate some of the ways in which different actors seek to mobilise various types of knowledge and evidence to shape energy systems and landscapes according to their own agenda. It also shows how claim-making makes use of discursive strategies that link energy development to wider social goals, and tactics like re-scaling that variously emphasise the local, regional and/or global dimensions of energy issues. In the final section of this chapter, we examine how these different forms of claim-making reveal underlying socio-political tensions at the root of energy conflicts.

ALTERNATIVE EXPLANATIONS AND RESPONSES TO ENERGY CONFLICTS

The ways in which claims are made about energy strategy and infrastructure siting reveal some of the key attributes of the conflicts that can emerge around current and future energy systems. However, as the example of wind energy illustrated, some of these conflicts are broader than arguments about the scientific merit of particular approaches or indeed the immediate economic and social costs. Rather, such conflicts are frequently about underlying differences in values and represent tensions in how society is governed and the dominant economic paradigm that has progressed energy policies in specific ways.

The case of wind energy (see Figure 7.4) provides a good illustration of the narratives that underpin different instances of claim-making. Devine-Wright (2005; 2009) provides an alternative narrative to what has commonly been referred to as NIMBYism or the 'Not in My Backyard' syndrome. Critically, NIMBYism is seen as demonstrating an inconsistency or selfishness in how people resist developments in their 'backyard' but are happy to see them developed elsewhere.

The use of the NIMBYism explanation to dismiss local opposition to renewable energy developments and other major projects has attracted significant critique over recent years, with a long line of commentators arguing that the term lacks academic credibility and a clear basis in social science theory and evidence. Accordingly, researchers have attempted to provide more sustained and sophisticated analyses of local reactions and opposition. Devine-Wright (2009) has proposed an alternative way of examining such reactions through the lens of social psychology, exploring ideas of place attachment and place identity.

NIMBY
the Not in My Back Yard (NIMBY) syndrome has commonly been used to explain the response of some in local communities to development projects that are perceived to have overtly negative impacts. NIMBYism is often used as a shorthand explanation for local opposition, with the suggestion that it is selfishly motivated and illegitimate rather than concerned with the greater good. However, use of the term has been widely criticised: protestors can have quite legitimate concerns about changes to where they live, are often concerned about wider issues not just very local ones, and may be objecting as much to how decisions are being made as to the nature of the development proposal itself.

Figure 7.4 An example of an offshore windfarm

Two key concepts from psychology are invoked: 'place attachment', which refers to the process of individuals becoming attached to a particular place, as well as the impact of this attachment; and 'place identity', which is related to the physical and symbolic attributes of a place and how they connect to and influences someone's sense of the self. Through using this theoretical perspective, Devine-Wright (2009) argues that major infrastructure and energy projects psychologic-ally disrupt both place attachment and place identity and, in so doing, a process of place disruption is enacted that involves becoming aware of a development, interpreting its potential impact on place, evaluating this impact, considering how to respond to change and eventually acting accordingly. In this way, social psychological interpretations of place attachment and place identity follow the tradition from social psycho-logical studies of focusing on individual responses to change.

A second approach to re-interpreting NIMBYism has taken a more sociological and normative perspective, exploring issues of democracy, participation and environmental justice. In contrast to the social-psychological approach described previously, research by authors such as Gross (2007) is interested in how major projects generate opposition through social conflicts that emerge between different groups, often as a result of the process of decision-making. Using a case study of a 69 turbine wind energy development in New South Wales, Australia, Gross (2007) highlights the discourses that emerged through decision making about the scheme, drawing on a range of semi-structured interviews. This work demonstrated that a lack of perceived fairness in the procedures used to determine the final outcome was critical, with residents claiming that their community had been 'divided' by the process, not least because it created 'winners' and 'losers' through playing-off different groups with the use of incentives and other strategies to obtain agreement. Gross (2007) argues that there was a clear sense of 'procedural injustice', which captures many of the challenges faced for the democratic process in reconciling competing interests. Indeed, given the different groups created by the process, Gross (2007) also argues that 'distributional' justice is a key feature. In this way, those analysing such developments need to be aware of the influence that decision-making processes have on perceived fairness and justice within given community, and how these different politics can have long-lasting social impacts. This was a theme also highlighted by Lesbriel and Shaw (2000), who argue that social scientists and practitioners need to find new ways of dealing with distributional justice: they advocate a shift from purely 'top-down' processes of energy development (which often involves forms of tokenistic consultation and participation) to **deliberative** forms of energy governance and embedded community engagement, which may involve aspects of **volunteering**. Such issues of procedure and voice run through many

deliberation
one of the key barriers for effective dialogue and discussion around energy conflicts has been the mechanisms conventionally deployed by energy developers for gaining public consent, which have often relied on consultation and top-down governance mechanisms. Researchers such as Burgess et al. (2007) have strongly argued for a more 'deliberative' approach to opening up knowledge controversies. Forms of deliberation vary, but the essence is captured in the goal of levelling 'knowledge hierarchies' between different groups, and using processes that enable debate and decision-making to be transparent and consensually accepted by all of those involved.

volunteering
incorporates deliberative approaches into a community-centred framework for decision-making about the siting of energy facilities. It is a process of active public engagement where knowledge about the costs, impacts and benefits of siting an energy facility are collaboratively derived with the affected community (i.e. co-production of knowledge), including potentially elements of citizen science, social learning and deliberation. The intention is that communities will feel empowered sufficiently to volunteer to host facilities, including deeply contentious ones such as nuclear waste sites (Bickerstaff 2012).

other examples of energy-related conflicts, including those framed as being about questions of climate justice.

Community-led renewable energy projects are an example of a direct challenge to existing power and knowledge hierarchies, promoting instead new forms of energy democracy (Kunze and Becker 2014). Taking forward such energy developments in a way that directly involves local people in decision-making and also distributes benefits locally (for example, through local cooperative ownership) rather than to distant corporate bodies and share owners, can be seen as a way of 'doing justice' and fundamentally shifting how energy governance is enacted at a local scale. Such projects do not necessarily avoid opposition (Walker et al. 2010; Simcock 2014) but are often seen as more likely to keep local people positively engaged and on-side (Walker and Cass 2007).

The ways in which energy developments create broader conflicts within and beyond communities and the need for a deeper social scientific understanding of how people react to such projects can also be set alongside some more fundamental responses to energy conflicts, many of which have been motivated by the need for greater energy security (see **Chapter 8**). Within developed economies, a recurring theme that has emerged is the issue of energy or fuel poverty (see **Chapter 5**). As Walker and Day (2012) have highlighted, fuel poverty relates to the affordability of key energy sources for domestic lighting, cooking and, most notably in a nation like the UK, heating. This issue is one that has received considerable political interest in recent years, in part because of economic austerity and declining relative incomes, but also because of the lack of effective regulation of energy companies and public outcry over large bonus payments for executives alongside rising domestic energy bills. Walker and Day (2012: 74) call for a fundamental change in the way that researchers and policymakers consider the needs and rights of vulnerable groups, which relates not only to the bounded nature of nation state energy policies, but also to the changing geographies of energy that are emerging in a world awakened to the need to de-carbonise:

> Fuel poverty interventions may be largely within the scale of the nation-state, but the implications of national energy policy spread beyond, tied to global resource flows and directly modifying the global commons of the atmosphere. In this way, as carbon mitigation measures impact on energy strategy and energy prices, the (mis)fortunes of the fuel poor are becoming increasingly folded into a climate justice framing.

Accordingly, we can view energy conflicts as ones that are layered and scaled. They represent apparently parochial and localised disputes about particular schemes or specific prices, but fundamentally, the research described in this

climate justice
a normative position that recognizes the causes and consequences of climate change are unequally distributed and seeks to redress this imbalance. Responsibility for emissions rest with more wealthy populations and carbon-intensive economies, particularly when historic emissions are taken into account, whereas the impacts of climate change are expected to be most severe for the poor and in the global South. Demanding justice in responses to climate change often focuses at an international level, but can also give attention to fairness within countries, including who pays for the costs of climate change mitigation.

energy democracy
shifting energy systems towards forms of local, community and municipal ownership and control has been interpreted as an assertion of energy democracy. Calls for energy democracy challenge established models of centralised, state or corporate owned and controlled energy infrastructures, and aim for much greater citizen participation in the development of sustainable, socially and environmentally just and inclusive energy transitions.

section highlights the inter-connections between specific energy disputes and broader issues of social, climate and energy inequalities.

CONCLUSION

For the reasons we have discussed in this chapter, conflicts over energy policies and development projects will undoubtedly endure. The diversity and intensity of consequences and impacts arising from the many different component parts of energy systems, and their uneven distribution socially and geographically, mean there are no simple solutions that can avoid disputation and disagreement. As we have argued, controversy and conflict are a necessary part of a functioning democratic system, and it is deeply problematic when dissent is suppressed by the state or corporate power – as it has been in a succession of energy-related controversies around the world. Competing claims over scientific knowledge, for rights to human health and ecological integrity, and for achiev- ing a balance between harnessing energy for social progress and limiting the negative consequences of doing so, need to be debated and subjected to scrutiny. Deliberative methods, engagement and participation in governance can help with the promotion of better forms of decision-making. But, as will be considered further in **Chapter 10**, there are also arguably intrinsic qualities to some forms of energy future that make them more amenable to democratic processes, and, it follows, less prone to sustaining the deeply embedded conflicts and inequalities that have characterised past and present energy system configurations.

QUESTIONS FOR DISCUSSION

• Why can conflict and controversy be seen as positive rather than negative features of energy system development?

• What examples of 'boundary work' around scientific claims related to energy issues can you identify, and what groups or organisations are involved?

• Why is environmental justice seen as a relevant concept for understanding energy conflict and controversy?

• What are the likely conflicts and social impacts that will emerge with shifting resource and energy geographies in the future?

ACTIVITIES/POTENTIAL RESEARCH PROJECTS

- Consider the causes and consequences of an energy conflict in your locality, such as a wind energy development, routing of pipelines or electricity pylons, or opposition to oil or gas exploration. What are the main arguments for and against such a development and who is making them? How do you assess the relative merits of their arguments and how do these impact on life in your locality or community?

- Consider how energy stresses will change in the future as a consequence of anthropogenic climate change and changes in the availability of fossil fuels: who will be impacted and how and why? Start locally – for example, if oil prices rise, what would be the impact on your neighbourhood? Consider both direct and indirect stresses.

RECOMMENDED READING

■ Bickerstaff, K. 2012. 'Because we've got history here': Nuclear waste, cooperative siting, and the relational geography of a complex issue. *Environment and Planning A* 44(11): 2611–2628.

The experience of a cooperative process of volunteering to host a radioactive waste disposal facility in West Cumbria, UK, is analysed in terms of the history, economy, and culture of the region and its intimate connections with the nuclear industry. At the time of writing it was the only area in which there were expressions of interest in hosting the facility. An analysis of the logics of volunteering in the context of experience of intense conflict and controversy of radioactive waste management is provided and how this became situated in a specific geographical context.

■ Devine-Wright, P. 2009. Rethinking NIMBYism: the role of place attachment and place identity in explaining place-protective action. *Journal of Community & Applied Social Psychology* 19: 426–441.

This paper critiques the NIMBY concept, which has been applied in a range of public debates concerning energy developments, most recently renewable energy infrastructures. In the paper, Devine-Wright provides a summary of key academic literature that has critiqued the overly simplistic NIMBY concept and uses his research on place and identity to argue for what he terms 'place protective' action, which is based on a series of stages in place-based settings, where processes of becoming aware, interpreting, evaluating, coping and acting are viewed at different scales, from the individual to the socio-cultural.

■ Leiserowitz, A.A., E. W. Maibach, C. Roser-Renouf, N. Smith and E. Dawson. 2013. Climategate, public opinion, and the loss of trust. *American Behavioral Scientist* 57: 818–837.

This paper explores the ways in which public trust in science can be understood from a psychological perspective, using the 2009 so-called 'Climategate' scandal as an example. Using data from representative samples of American participants in 2008 and 2010, the paper argues that beliefs in the science of anthropogenic climate change were severely dented by the release of emails from climate scientists in the United Kingdom and United States. However, the authors argue that reductions in levels of trust were most significant amongst those with an individualistic worldview and overall, trust in scientists still remained higher amongst Americans than for other groups.

■ Schaeffer, C. and M. Smits. 2015. From matter of facts to places of concern? Energy, environmental movements and place-making in Chile and Thailand. *Geoforum* 65: 146–157.

This paper provides an analysis of how energy turns into a "matter of concern", tracing how in energy controversies environmental movements frequently play a central role in making claims about the consequences that will follow from development projects. Through an analysis of case studies in Chile and Thailand it is argued that energy controversies turn villages, cities, or regions themselves into "places of concern", challenging and disrupting established senses of place and place values.

REFERENCES

Barnett, J. and J. Campbell. 2009. *Climate change and small island states: power, knowledge and the South Pacific*. London: Earthscan.

Barr, S. 2011. Climate forums: virtual discourses on climate change and the sustainable lifestyle. *Area* 43(1): 14–22.

Bickerstaff, K. 2012. 'Because we've got history here': nuclear waste, cooperative siting, and the relational geography of a complex issue. *Environment and Planning A* 44(11): 2611–2628.

Bickerstaff, K., G. Walker and H. Bulkeley. (eds) 2013. *Energy justice in a changing climate: social equity and low carbon energy*. London: Zed.

Bomberg, E. 2017a. Shale we drill? Discourse dynamics in UK fracking debates. *Journal of Environmental Policy & Planning* 19(1): 72–88.

Bomberg, E. 2017b. Fracking and framing in transatlantic perspective: a comparison of shale politics in the US and European Union. *Journal of Transatlantic Studies* 15: 101–120.

Boucher, P. 2012. The role of controversy, regulation and engineering in UK biofuel development. *Energy Policy* 42: 148–154.

Breen, J., S. Dosemagen, J. Warren and M. Lippincott. 2015. Mapping grassroots: geodata and the structure of community-led open environmental science. *Acme-an International E-Journal for Critical Geographies* 14(3): 849–873.

Burgess, J., A. Stirling, J. Clark, G. Davies, M.Eames, K. Staley and S. Williamson. 2007. Deliberative mapping: a novel analytic-deliberative methodology to support contested science-policy decisions. *Public Understanding of Science* 16: 299–322.

Cart, J. 2010. Tiny group has big impact on spill estimates. *Los Angeles Times*, 1 May 2010.

Carvalho, A. 2007. Ideological cultures and media discourses on scientific knowledge: re-reading news on climate change. *Public Understanding of Science* 16: 223–243.

Cotton, M. 2016. Fair fracking? Ethics and environmental justice in United Kingdom shale gas policy and planning. *Local Environment*. Available online at http://dx.doi.org/10.1080/13549839.2016.1186613

Devine-Wright, P. 2005. Beyond NIMBYism: towards an integrated framework for understanding public perceptions of wind energy. *Wind Energy* 8: 125–139.

Devine-Wright, P. 2009. Rethinking NIMBYism: the role of place attachment and place identity in explaining place-protective action. *Journal of Community & Applied Social Psychology* 19: 426–441.

Devine-Wright, P. and H. Devine-Wright. 2006. Social representations of intermittency and the shaping of public support for wind energy in the UK. *International Journal of Global Energy Issues* 25(3): 243–256.

Gieryn, T. 1983. Boundary-work and the demarcation of science from non-science: strains and interests in professional ideologies of scientists. *American Sociological Review* 48: 781–795.

Gieryn, T. 1999. *Cultural boundaries of science: credibility on the line.* Chicago, IL: University of Chicago Press.

Gross, C. 2007. Community perspectives of wind energy in Australia: the application of a justice and community fairness framework to increase social acceptance. *Energy Policy* 35: 2727–2736.

Intergovernmental Panel on Climate Change (IPCC) 2014. *Climate change 2014: impacts, adaptation, and vulnerability.* Geneva: IPCC.

Jackson, A.L. 2011. Renewable energy vs. biodiversity: policy conflicts and the future of nature conservation. *Global Environmental Change* 21(4): 1195–1208.

Kunze, C. and S. Becker. 2014. *Energy democracy in Europe: a survey and outlook.* Brussels: Rosa Luxemburg Stiftung.

Kurtz, H.E. 2002. The politics of environmental justice as the politics of scale: St James Parish, Louisiana and the Shintech siting controversy. In *geographies of power: placing scale.* A. Herod and M.Wright (eds), Oxford: Blackwell, pp. 249–273.

Leiserowitz, A.A., E.W. Maibach, C.Roser-Renouf, N. Smith and E. Dawson. 2013. Climategate, public opinion, and the loss of trust. *American Behavioral Scientist* 57, 818–837.

Levin, S. 2016. Dakota pipeline protesters set for 'last stand' on banks of Missouri river. Available online at www.theguardian.com/us-news/2016/oct/31/north-dakota-access-pipeline-protest-last-stand

Levy, B. and J. Patz. 2015. *Climate change and public health.* Oxford: Oxford University Press.

Leiserowitz, A.A., E.W. Maibach, C. Roser-Renouf, N. Smith and E. Dawson, E. 2013. Climategate, public opinion, and the loss of trust. *American Behavioural Scientist* 57(6): 818–837.

Lesbirel, S.H. and D. Shaw. 2000. Facility siting: issues and perspectives. *Challenges and issues in facility siting.* Columbia Earthscape. Available online at www.researchgate.net/profile/Daigee_Shaw/publication/268033474_FACILITY_SITING_ISSUES_AND_PERSPECTIVES/links/55d3c9ee08ae7fb244f58f5c.pdf

Mail on Sunday. 2013. Finally, the hard proof that shows global warming forecasts costing you billions were all wrong. *Mail on Sunday*, 17 March 2013.

Matejek, S. and T. Gossling. 2014. Beyond legitimacy: a case study in BP's 'green lashing', *Journal of Business Ethics* 120(4): 571–584.

Mouffe, C. 1998. *Hearts, minds and radical democracy.* Red Pepper. Available online at www.redpepper.org.uk/hearts-minds-and-radical-democracy/

Naus, J., G. Spaargaren, B.J.M. van Vliet, M. Hilje and D. van der Horst. 2014. Smart grids, information flows and emerging domestic energy practices. *Energy Policy* 68: 436–446.

Nixon, R. 2011. *Slow violence and the environmentalism of the poor.* Cambridge: Harvard University Press.

Omerje, K. 2005. Oil conflict in Nigeria: contending issues and perspectives of the local Niger Delta people. *New Political Economy* 10(3): 321–334.

Palmer, J.R. 2014. Biofuels and the politics of land-use change: tracing the interactions of discourse and place in European policy making. *Environment and Planning A* 46(2): 337–352.

Pidgeon, N., I. Lorenzoni and W. Poortinga. 2008. Climate change or nuclear power – No thanks! A quantitative study of public perceptions and risk framing in Britain. *Global Environmental Change* 18(1): 69–85.

Roberts, T. 2013. Energy siting governance: social and political challenges associated with the development of low-carbon energy in marine environments. T. Roberts, P. Upham, R. Thomas, S. Mander, C. McLachlan, P. Boucher, C. Gough and D.A. Ghanem (eds), *Low-carbon energy controversies.* Abingdon and New York: Routledge, pp. 114–131.

Safford, T. G., J.D. Ulrich and L.C. Hamilton. 2012. Public perceptions of the response to the Deepwater Horizon oil spill: personal experiences, information sources, and social context. *Journal of Environmental Management* 113: 31–39.

Schaeffer, C. and M. Smits. 2015. From matter of facts to places of concern? Energy, environmental movements and place-making in Chile and Thailand. *Geoforum* 65: 146–157.

Simcock, N. 2014. Exploring how stakeholders in two community wind projects use a 'those affected' principle to evaluate the fairness of each project's spatial boundary. *Local Environment* 19(3): 245–258.

Sovacool, B.K. 2008. Valuing the greenhouse gas emissions from nuclear power: a critical survey. *Energy Policy* 36(8): 2950–2963.

Stanley, A. 2008. Citizenship and the production of landscape and knowledge in contemporary Canadian nuclear fuel waste management. *The Canadian Geographer* 52(1): 64–82.

Stoll-Kleemann, S., T. O'Riordan and C.C. Jaeger. 2001. The psychology of denial concerning climate mitigation measures: evidence from Swiss focus groups. *Global Environmental Change* 11(2): 107–117.

Van der Horst, D. and D. Toke. 2009. Exploring the landscape of wind farm developments; local area characteristics and planning process outcomes in rural England. *Land Use Policy* 27(2): 214–221.

Walker, G. 2012. *Environmental justice: concepts, evidence and politics.* Abingdon: Routledge.

Walker, G. and G. Baigorrotegui. 2016. Energycoop Aysen. In *Cultures of community energy: international case studies.* N. Simcock, R. Willis and P. Capener (eds), London: The British Academy, pp. 65–68.

Walker, G. and N. Cass. 2007. Carbon reduction, 'the public' and renewable energy: engaging with sociotechnical configurations. *Area* 39(4): 458–469.

Walker, G., S. Hunter, P. Devine-Wright, B. Evans and H. High. 2010. Trust and community: exploring the meanings, contexts and dynamics of community renewable energy. *Energy Policy* 38: 2655–2633.

Walker, G. and R. Day. 2012. Fuel poverty as injustice: integrating distribution, recognition and procedure in the struggle for affordable warmth. *Energy Policy* 49: 69–75.

Walker, G., P. Devine-Wright, J. Barnett, K. Burningham, N. Cass, H. Devine-Wright, G. Speller, J. Barton, B. Evans, Y Heath, D. Infield, J. Parks and K. Theobald. 2011. Symmetries, expectations, dynamics and contexts: a framework for understanding public engagement with renewable energy projects. In *Renewable energy and the public: from NIMBY to participation.* P. Devine-Wright (ed.), London: Earthscan, pp. 1–14.

Walker, R. 2011. The impact of Brazilian biofuel production on Amazônia, *annals of the Association of American Geographers* 101(4): 929–938.

Watts, M. 2004. Resource curse? Governmentality, oil and power in the Niger Delta, Nigeria. *Geopolitics* 9 (1): 50–80.

Whitmarsh, L., 2009. What's in a name? Commonalities and differences in public understanding of "climate change" and "global warming". *Public Understanding of Science* 18(4): 401–420.

Williams, D. and K. Baverstock. 2006. Chernobyl and the future: too soon for a final diagnosis, *Nature* 440: 993.

Wittneben, B. 2012. The impact of the Fukushima nuclear accident on European energy policy. *Environmental Science and Policy* 15(1): 1–3.

Energy securities

- Understand the concept of energy security and its main constitutive elements.
- Appreciate the different practices associated with attempts to make energy secure throughout the supply chain, and at a range of geographical scales.
- Explore the limits of conventional accounts of energy security by asking for whom energy is made secure.
- Examine some of the different ways in which energy can become insecure at various material sites.

This chapter introduces and critically evaluates the notion of 'energy security'. Concerns surrounding the security of energy supplies are at the forefront of media attention and policy debate in many parts of the global North and South. The prominence of energy security as a policy concern reflects the nature of the contemporary global economy and the energy systems on which it depends: prevailing models of economic growth require abundant, uninterrupted inputs of energy, with much of this supplied by transnational production networks and an international trade in fossil fuels such as oil and gas. The extension of energy production chains across vast distances means that sites of energy consumption are often located in different political jurisdictions to the locations of resource supply. The dependencies and vulnerabilities this creates lies behind discussions of 'energy independence' in countries like the United States, Poland and South Korea, which have relied heavily on imported hydrocarbons. Policies aimed at addressing the energy security of countries with limited domestic resources emphasise the need to diversify forms of energy supply within their political boundaries, reduce energy demand, and/or establish diplomatic and economic networks that reduce the risk of interruption to key energy chains.

As a whole, mainstream deployments of the energy security concept in government policy and the media privilege some material sites and horizons of action over others. Energy security is primarily seen as a problem of cross-border resource supply that affects the scale of the nation state (**Chapter 4**). Yet the nexus of security and energy is not limited to the nation state and operates at other levels too. Households, neighbourhoods, towns and regions can also be affected by short- or long-term disruptions to the physical supply and/or cost of energy. In the same way, practices of securing energy often extend beyond formal state actors to encompass the private sector, and a complex array of political and economic strategies.

This chapter explores what is at stake in the securitisation of energy – i.e. the turning of energy into a security concern (as distinct from, say, a matter of environmental sustainability, poverty alleviation, or democratic control). To do this it contextualises the role energy security currently plays in political debate, and provides an alternative reading that examines some of the complexities surrounding the concept and its application. This chapter highlights some of the different ways in which energy is secured in practice, from strategic planning contexts all the way down to the actions of individual households in securing their own supplies via, for example, a range of specific incentives to promote insulation and micro energy generation. Finally, the chapter scrutinises the causes and impacts of energy insecurity, adopting several different scales of analysis from international examples to notions of individual energy scarcity. Through these examples, we consider the ways in which people and places can become vulnerable, not least because of the embeddedness of energy in different economic and social practices. Throughout, we emphasise the value of thinking about energy security outside a conventional framework limited to national energy supply, and of asking *for whom* energy is being made secure.

> **energy securitisation**
> describes the process whereby particular groups develop the level of energy security they desire and maintaining this over time. A critical approach to securitisation centres on the question 'for whom?' and explores the strategies deployed by individuals, place-based communities and nations to achieve their desired level of energy security. Such processes can be political in nature and involve prioritising one group in society over another; they can also be economically and spatially defined.

SECURITISING ENERGY

The security of energy supply is now a central policy concern in many countries. The linking of energy with security may seem obvious but it is not trivial: it elevates energy as a priority risk for governments and a matter of strategic concern, placing it on a par with other threats to national and international security such as terrorism, cyber-attacks and organised crime. Once energy is framed as a matter of national security, a broad range of energy actions and interventions can be justified in the name of securing the country from threat, from domestic energy resource development and large-scale energy infra-structure projects, to by-passing or abbreviating processes of public consultation and curtailing rights to protest, unionise or strike (Bridge 2015).

Much of the mainstream policy interest in energy security hinges on the continuing availability – in both the short and long term, of fossil fuels responsible for powering key sectors of the economy. Sectors like transport, housing and industry are particularly vulnerable to interruptions in supply.

Short-term disruptions to imports of oil and natural gas, for example, can lead to severe disturbances in the normal conduct of economic and social life. The impact of this kind of disruption was forcefully demonstrated in the 1970s, when two major geo-political events created instability in the oil market and led to sharp increases in the price of oil. Contemporary policy interest in energy security originates from this period, as do many popular discourses about the value of energy independence. In the first of the 1970s 'oil shocks', the Organisation of Arab Petroleum Exporting Countries (OAPEC) declared an embargo on oil shipments to North America, Japan and parts of Europe as a result of the United States' support of Israel in the Yom Kippur War. This sent global oil prices sharply upwards (from US$2 to $10 per barrel) and petroleum rationing was introduced in a number of countries, including the United States. Many motorists experienced physical shortages of fuel in this period. Prices rose sharply again in 1979 as a result of the Iranian Revolution, as oil output in the Middle East fell and market panic ensued at the prospect of major shortages. These two formative events continue to shape how energy security is conventionally understood: as security against the threat of disruption to the international supply network for fossil fuels (Boyle et al. 2003).

There is also the prospect of limits on the availability of coal, oil and gas in the longer term. These may arise from physical supply-side constraints, such as the 'peaking' of conventional oil output. The 'peak oil' argument asserts there are geological constraints on the ability to continuously expand physical supplies of conventional crude oil. Moreover, it holds that this inevitable maximum point of crude oil production is imminent (or indeed, may have passed) and that output of conventional crude oil will enter a terminal state of decline. To many 'peak oilers', the decline in crude production is not something to be welcomed on environmental grounds but a source of concern: rising prices and physical oil shortages will, they argue, exacerbate economic and political tensions and, at the extreme, trigger an apocalyptic scenario of social collapse. However, such predictions have yet to materialise. Rapid growth in the output of unconventional sources of oil (**Chapter 4**) has offset any peak in the production of conventional crude. At the same time, there is growing evidence that a peak in global oil production will come not from shortages of supply, but from reductions in demand for oil driven, in part, by action on climate change. Nonetheless, peak oil and the prospect of physically induced shortages has played a significant role in galvanising support for energy security policies in a number of countries.

Limits on the physical availability of fossil fuels may arise from policy action to address climate change, particularly those that focus on 'shutting in' fossil fuels – i.e. not extracting known reserves of coal, oil and gas. Such supply-side actions are currently at the margins of climate change mitigation which, in the main, targets consumption and demand. There are notable moves in this direction, however, that suggest a 'carbon constraint' may be emerging on the availability of fossil fuels. The Chinese government's announcement (2017) of a moratorium on new coal-fired coal plants, and its closure of inefficient coal mines, is indicative of this kind of supply-side action (although reasons for these actions are not solely related to air quality or climate change). Moves to suppress demand for carbon-intensive fuels, and the growing prospect of climate-driven constraints on their supply, mean that industrial societies urgently need to find alternative methods of procuring the dominant share of total primary energy currently provided by oil, coal and gas.

Alongside the geopolitical and physical threats to supply described above, energy security concerns also centre on issues of investment and a number of specific technological challenges. The reliability of supply can be compromised by insufficient investment in resource development, generation capacity and transmission and storage infrastructure. The provision of spare capacity and diversity of supply can also enhance energy security (Bahgat 2006). The difficulty of efficiently storing large quantities of electrical energy is a specific technological concern. Many consumers have come to expect electricity to be available at the flick of a switch – i.e. whenever and wherever it is needed. A priority for businesses and governments, therefore, is to secure a constant and reliable supply of electricity, regardless of short- and long-term fluctuations in overall demand. The flow and storage properties of electricity, therefore, are regularly a feature of debates surrounding future threats to energy security. Popular media frequently invoke the possibility of the 'lights going out' to highlight inadequate levels of investment in supply or poor infrastructure management. The social, economic and technological consequences of electricity blackouts and involuntary load-shedding – and of brownouts caused by voltage reductions to reduce the load on electrical systems – have been central to efforts in many countries to highlight the need and urgency of addressing energy security.

Increasingly, the securitisation of energy as a policy concern is linked to climate change. Energy security is both a cause of climate change (e.g. the commitment of many state and corporate actors to securing energy systems in which fossil fuels play a dominant role), and a consequence of it (e.g. the desire to ensure future energy supply in the face of large environmental uncertainties arising from climate change). Policies aimed at reducing greenhouse gas emissions have focused on the need to move away from the predominance of fossil fuels in the supply mix, onto renewable and low carbon forms of energy. However, electricity cannot be supplied by renewable resources under the same socio-technical conditions that have characterised its generation from hydrocarbons. This is principally due to the lower energy density of renewables, which means that larger surface areas are required for the generation of electricity when compared with fossil fuels (see **Chapter 1**). Renewables are also temporally intermittent, and the flow of energy from them is variable as a result of factors outside the direct control of the producers of energy. Their production is spatially diffuse and can often involve multiple small-scale providers distributed across wide geographical areas. Thus, new infrastructures for electricity backup, storage and transmission are required in order to ensure a stable and reliable supply from such resources. Furthermore, climate change will have a direct impact on energy supplies as a result of physical and social changes to the earth's environment. These include reduced water resource availability which, in regions where hydropower is a significant part of the energy mix, will impact the production of electricity from hydro-electric sources (such as the recent spate of seasonal droughts in Brazil); as well as short- and long-term alterations in patterns of energy demand (e.g. for cooling) as a result of climate change.

As a result of these different contingencies – geopolitical, physical, techno-logical – energy security has come to be understood primarily as a question of energy supply, and as a strategic matter for national governments. Indeed, 'securing energy' has been an important rationale in forging international political networks and geopolitical relations in many parts of the world. For example, the International Energy Agency – perhaps today the pre-eminent multilateral institution for energy governance – was established by the OECD in 1974 to enable strategic co-operation by member countries on the security of oil supply, following supply reductions from the Middle East. Likewise, the European Union is consolidating a common energy policy approach and developing infrastructure projects that enhance energy security across the territory of its 28 Member States: the EU's goal is to create a space in which infrastructure, import and storage capacities can be shared, so as to decrease the potential for supply interruptions. Geopolitical competition in East Asia to secure energy supplies and regional influence lies behind territorial disputes in the East China and South China seas. Elsewhere, concerns over energy security have played a crucial role in shaping support for domestic hydrocarbon production, including unconventional gas development in the US, UK, Australia and South Africa, and new coal mines in Poland and Turkey.

EXPANDING DEFINITIONS OF NATIONAL ENERGY SECURITY

The meaning of energy security has expanded over time so that, by one estimate, there are now over 40 different definitions (Sovacool 2013). Such a broad range of understandings reflects, in part, the way a discourse of 'securing energy' has emerged around several separate policy agendas. Three perspectives, in particular, stand out (Cherp and Jewell 2011): a sovereignty perspective, centred on geopolitical risks and the need to protect flows across national borders; a view of energy security as the 'robustness' of infrastructure, in terms of its capacity to handle growing demand or cope with extreme natural events; and an understanding of security as resilience in the face of unpredictable, complex and coupled social and natural systems. Of these, it is the sovereignty perspective that dominates mainstream accounts of energy security. A central concern has been to ensure the physical availability of oil and gas imports, notwithstanding the rising prominence of electricity in the global supply mix. Attention has focused on geopolitical relations associated with the international energy trade, particularly with respect to the need to manage oil markets. A priority has been to ensure uninterrupted access to key energy carriers (notably crude oil and related fuels), with strategies directed towards diversifying supply in terms of both the mix of energy carriers involved and the geographies of energy supply. Strategic response has also centred on embedding **redundancy and liquidity** within energy systems, enabling them to withstand external shocks, and the safeguarding of self-sufficiency in energy provision (Chester 2010).

Mainstream understandings of energy security have frequently highlighted the central role of markets to energy supply. This means that a steady and reliable physical supply of energy is not the only criterion in determining a country or region's level of energy security: the price at which energy supplies are available also matters. This combination of *availability* and *affordability* is at the core of most traditional definitions of energy security, such as the "availability of energy at all times in various forms in sufficient quantities and at affordable prices" (Umbach 2004: 142). The International Energy Agency offers a similar yet more succinct definition: the "availability of a regular supply of energy at a reasonable price". Approaches to energy security that emphasise the importance of price and affordability point out that the liberalisation of energy markets (**Chapter 2**) means energy security is now increasingly determined by market forces. Accordingly, mechanisms for matching energy supply with demand

redundancy and liquidity are seen as universal principles of energy security at the national scale. The former refers to the installation of spare capacity in energy systems, and can include the expansion of import infrastructure, the provision of additional generation and transmission network capacity, as well as the construction of supplementary storage facilities. The latter refers to the development of competitive markets in which energy can be traded with as little as possible regulatory or political impediment.

EXAMPLE 1: UNDERSTANDING ENERGY INSECURITY AS A STRUCTURAL CHALLENGE

The UK government estimates that one in ten households in England lived in fuel poverty in 2014, equivalent to over 2.3 million households (Department for Energy and Climate Change, 2016). Fuel poverty is likely to be greater in older properties, in buildings constructed with solid walls, in properties where no boiler is present and where there is no access to the gas grid. As such, fuel poverty is not simply a function of an individual ability to pay for fuel: it is also, importantly, a function of the building stock in which people dwell. This argument is made by persuasively by the European economics consultancy group, Frontier Economics (2015): re-conceptualising energy efficiency as an infrastructural issue, they argue, recognises the critical role that domestic spaces play in influencing fuel poverty, and the health and well-being impacts of cold, damp living conditions. A systematic approach to 'building in' energy efficiency as national infrastructure, they suggest, would have quantifiable benefits comparable to the construction of a second national high speed rail line, even without accounting for the health and well-being benefits. Such arguments critically re-position the issue of fuel poverty away from a focus on individual 'ability to pay' arguments, towards a recognition of the structural injustices faced by those in fuel poverty (Walker and Day 2012).

are crucial for the maintenance of energy security. Proponents of market-based forms of energy security argue that competitive, well-functioning markets supported by independent regulation are key. Others, however, take a different view and claim that government intervention is needed because markets alone "fail to ensure adequate and sustained levels energy supply security" (Gnansounou 2008: 3734).

In certain contexts, the 'sovereignty' perspective at the heart of mainstream accounts of energy security has centred on the maintenance of energy *demand* rather than supply. Ensuring reliable and dependable energy demand is important for countries whose energy exports have key role in supporting industrial sectors and state budgets, and in maintaining the 'social contract' between ruling governments and their populations. For example, oil exports account for around 40% of Saudi Arabia's GDP, around 85% of export earnings and most government revenue. The country's public finances, therefore, are tightly linked to oil exports. A similar relationship exists in Turkmenistan in relation to gas exports (primarily to China), with gas representing around 30% of the country's GDP and over half its industrial output; in Qatar, where oil and gas account for around 60% of GDP, 70% of government revenue, and over 80% of export earnings; and in Russia, where oil and gas represent around two-thirds of export revenues. With public finances so reliant on energy exports, governments in these countries are concerned to secure reliable, long-term demand for their hydrocarbon products. They are frequently keen to reduce price volatility and preserve prices within a range: low enough to not substantially weaken long-term demand (via substitution and/or efficiency gains), but high enough to achieve national revenue goals. Security of *price* is particularly significant for states that have extended government spending on the back of resource revenues, and which accordingly have seen the 'break-even' price (of oil and gas) necessary to finance annual state expenditure rise over time (Bridge 2015). For major energy exporters like Saudi Arabia, Venezuela or Qatar, prolonged reductions in revenue flows linked to oil and gas exports can have significant consequences for regime stability. They seek to secure energy demand by extending long-term supply contracts and protecting physical export routes, an interest they share with energy-consuming countries. Securing market share over the short to medium term is a typical concern of energy exporters – i.e. competition with other suppliers. But there is a growing concern too with the prospect of longer-term 'demand destruction', particularly for oil, through a combination of improvements in energy efficiency, fuel shifting in transportation (e.g. the rise of electric vehicles), rapid deployment of renewables, and the impacts of air quality and climate regulation. In this context, energy security strategy extends to policies of economic diversification that over time reduce a country's dependence on hydrocarbon exports. For example, Saudi Arabia's 'Vision 2030', launched in 2016, aims to reduce the country's dependency on oil via investment in the non-oil economy, the growth of non-oil trade and revenue, and building up a renewable energy sector.

The 4 As: availability, affordability, accessibility and acceptability

A useful framework for thinking about energy security expands beyond the core concerns of availability and affordability described above. The so-called '4 As' approach also includes the accessibility and acceptability of energy supply options. Accessibility here refers to the geopolitical aspects associated with gaining access to energy resources, while acceptability refers to the social and environmental effects of energy production (Kruyt et al. 2009). Concerns about the environmental consequences of energy production, and the exclusionary and social justice effects of rising energy prices, have become increasingly central within debates over energy policy. An expanded definition of energy security, incorporating one or both of these additional issues, is increasingly recognised by key energy governance institutions such as the IEA (2017), which defines energy security as "the uninterrupted physical availability at a price which is affordable, while respecting environment concerns". Affordability and acceptability are interesting criteria because they change the logic of energy security from one of raw survival (what Ciută (2010) calls a 'logic of war') to include profoundly moral and ethical questions about human welfare and sustainable development. They broaden the meaning of energy security from a traditional quantitative concern with the rate of supply (barrels of oil/cubic meters of gas per day) to more qualitative assessments of the socio-economic and environmental consequences of these flows (Bridge 2015). In a subtle yet very significant way, they extend energy security to consider the relationship of supply to prevailing practices and expectations around the consumption of energy services (e.g. mobility, heating, power and light).

Greater attention to who receives energy services, and who bears the social and environmental costs of securing them, promotes a view of energy security as resilience rather than a matter of sovereignty (Cherp and Jewell 2011). It directs strategy away from managing perceived geopolitical threats to supply, and towards improving the flexibility and diversity of energy systems and their ability to adapt in the face of both supply uncertainties and multiple social demands. In this way, expanding the definition of energy security to consider who is able to receive energy services and at what (economic, social and environment) cost brings traditional energy security concerns closer to questions of equity, justice and power which lie at the heart of the contemporary energy challenge (see **Introduction**). At a time when energy is frequently 'securitised' in ways that make maintaining supply synonymous with the fate of the nation, important questions to ask are security for whom, for which values, and from what threats? (Cherp and Jewell 2014: 415).

Managing energy security through indicators

Policy interest in energy security has encouraged the development of quantitative measures and indicators through assessments of variation in security (across

reserve-production ratio
the physical availability of a given energy source as it relates to depletion rate is commonly invoked in the context of energy security debates. Hence, the reserve-production ratio: a number, expressed in years, which quantifies the remaining stock. Globally, the current R/P ratio for coal is estimated at more than 150 years, while those for oil and natural gas range between 60 and 80. Care has to be taken in interpreting R/P ratios as they do not take into account reserve growth via new discoveries and changes in technology and/or economic conditions: the global R/P ratio for crude oil, for example, has remained pretty much the same for the last couple of decades despite a large expansion in production in that time.

import dependency
the situation a nation or region reaches where demand for energy exceeds the supply available (either materially or economically) from within that nation or region. Import dependency is often regarded as a holistic phenomenon for 'energy' overall, but there are frequently variations among energy sources according to the rate of discovery of new sources, market price changes and political instability.

reserves and resources
the distinction between reserves and resources is fundamental to contemporary understandings of energy security. While resources describe the entire physical stock of a given source, reserves refer to quantities that can be expected to be extracted under existing economic and technological conditions. In the case of oil, reserves can further be distinguished by the

space and time) can be made. The UK government, for example, publishes a 'Statutory Security of Supply Report' that conveys to Parliament important information about "the availability of electricity and gas to meet the reasonable demands of consumers in Great Britain". The development of an 'oil vulnerability index' also follows this trend. The index is made up of metrics such as the ratio of oil imports to gross domestic product, oil consumption per unit of GDP, and the share of oil in total energy supply. It also includes various 'risk indicators', such as the ratio of domestic reserves to oil consumption, exposure to geopolitical oil market concentration risks, the diversification of supply sources, and levels of political risk in oil-supplying countries (Gupta 2008). As this example suggests, energy security indicators commonly rely on metrics about physical availability, such as the **reserve-production ratio**, which emphasises the size of remaining natural resource stocks in relation to existing production rates; and about extent to which consumption is met by domestic supply, such as **import dependency**. How governments estimate the extent to which a **resource** can be converted into useful **reserves** is also of crucial importance in this context (Figure 8.1).

A major implication of developing and using energy security indicators is that a country can actively manage its level of energy (in)security using socio-technical procedures. The '4 Rs' of energy security (Hughes 2009) is a framework for managing insecurity that involves four steps: Review (understand the problem in terms of existing and future sources, suppliers and services, and the state of infrastructure and energy intensity); Reduce (use less energy via conservation and efficiency); Replace (shift to sources that are deemed more secure via fuel switching and diversification, changing suppliers or introducing new infrastructures); and Restrict (limiting additional demand to sources that are deemed secure). Managerial approaches like these, however, are based on specific assumptions about spatial and technical organisation, such as the connectivity of infrastructure, that are necessary to achieve desired levels of energy security for national economies. An underlying premise is that energy is supplied and managed by large, centralised utility companies that are regulated by governments and can access global energy markets. There is little space in such paradigms for understanding the provision of energy security via multiple small-scale and diffused sources.

The conceptual limitations of existing energy security approaches have been increasingly challenged by authors who emphasise the need for securing energy flows throughout the supply chain, and below the scale of the nation state. These alternative perspectives involve a shift in focus, away from viewing the energy system as something with the capacity to generate and enhance security for the nation, and towards thinking about the way energy systems are themselves exposed to

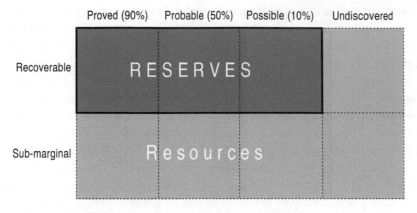

Figure 8.1 The distinction between energy reserves and resources

internal and external risks and threats. Threats to energy provision can come in the form of short-term, direct shocks caused by inadequate production capacity, technical or operational errors, and sudden episodes such as terrorist attacks or weather events; and longer-term, indirect stresses resulting from inadequate investment in new capacity and quality, poor maintenance of existing systems, lack of physical protection and political unrest. Issues of political trust, institutional regulation and the nature of international relations underpin both shocks and stresses (Johansson 2013). The duration, severity and location of energy supply disruption can produce very different consequences. Rapid events occurring closer to the end of the energy chain – i.e. near the point of final consumption – are more likely to create severe social consequences, in comparison to longer-term pressures occurring closer to sources of supply. For example, a sudden electricity blackout can cause widespread economic and social disorder, while a more sustained oil supply crisis might not be immediately felt by the consumers but can trigger significant realignments in state policy.

probability of recovery: if this probability is higher than 90% they are termed 'proven', while the terms 'probable' and 'possible' correspond to equivalent percentages of at least 50 and 10. Initial estimates of reserves tend to be conservative, subsequently expanding in size as confidence in the level of recovery and the quality of measurement methods increase.

ALTERNATIVE SCALES OF ENERGY (IN)SECURITY

Just as nation states have attempted to secure energy through a range of policy and economic instruments, so too have cities, towns and other communities reacted to concerns over rising energy costs and supply disruption in an effort to ensure energy service provision (Allen et al. 2012). These responses can broadly be categorised into those that are institutional in nature (led by or supported from the local state) and ones that lie 'below' or 'beyond' the state. The latter offer a creative space in which different energy landscapes can emerge, particularly in communities that are off-grid (Walker and Devine-Wright 2008).

Many of the new forms of institutionalised energy security have emerged at the urban scale. Cities and towns have taken a lead in developing new approaches, involving local government, developers and non-governmental organisations and often cast within the context of mitigating climate change. These range from technological modifications within conventional supplier-consumer relationships, to imaginative ways of communalising supply and demand. Three main approaches have emerged. First, urban areas have adopted a diversity-based approach for meeting energy needs, on the grounds that diversity in supply affords some protection against price and supply volatility, and ensures that broader climate change emissions targets can be attained. An example of this approach is the way in which many European cities have diversified public transport networks to rely on a mix of energy supplies, from conventional hydrocarbon based fuels (diesel, petroleum) to natural gas, biofuel and electricity. This is reflected in different transport modes, that, integrated with sustainable urban planning, aim to work towards what Banister (2008) has termed a 'sustainable mobility paradigm': transport systems that promote sustainable development by reducing the need for travel, and by framing mobility as a set of collective, rather than individual, acts of movement. A practical example of this approach can be seen in the Spanish city of Seville, which has developed a diversified public-transport strategy. This incorporates several different ways of getting around, including electric tram, natural gas fuelled buses, a public hire cycle scheme and walking, all facilitated by re-modelling much of the city centre. In this way, the city has sought to simultaneously broaden its energy supply base (with a greater focus on the renewables potential afforded by photovoltaic power) and achieve a net reduction in energy use for transport overall (MetroSeville 2014).

This first approach to generating greater energy security relies on a strategic planning framework and rests largely on local government and associated service providers. A second approach requires harnessing the different and often competing goals of local states, developers and local interest groups on a site-by-site basis. In many countries, the goal of decarbonising domestic energy supply has enabled planners to insist that housing developments have a range of renewables incorporated into their design, alongside measures to reduce energy use through insulation and other measures. On the one hand, this has been achieved through small-scale showcase developments, partly financed by local government. The Oak Meadow affordable housing scheme in North Devon, the UK, is an example of this process, where a range of measures have been adopted to reduce energy consumption and create healthy living spaces for tenants (Gale and Snowden Architects 2014). An alternative and more radical approach is community energy generation. This challenges the accepted wisdom that individual households should secure their own energy services via, for example, including a central heating boiler in every home. Community energy schemes take a wide variety of forms and not all focus on energy production. The most widespread, however, are District Heating Systems, which heat a number of homes within a defined area (often using a renewable energy source like wood chips, for example). Such systems offer efficiency savings when compared with individual household boilers, and there are other benefits too: reduced carbon emissions, space savings from the absence of chimney flues and other associated fixtures within individual homes, and ultimately an energy source that is comparatively reliable (see **Chapter 3**).

A final institutionalised form of energy security adopted in many contexts centres on individual households, and has been enabled by recent progress in micro-generation technologies like photovoltaic cells and geo-thermal energy (Mendoca 2007). In many parts of Europe, individual households have recently been incentivised by national governments to install micro-generation (DCLG 2013), alongside more longstanding subsidies for insulating homes and improving energy efficiency (**Chapter 3**). The value and significance of such schemes have been keenly debated. In practice, they frequently target socio-economic groups that are the most able to make required investment, raising issues of energy justice (Del Rio and Gual 2006).

Beyond the actions of national and local state governments, the goal of energy security has also been pursued by communities acting outside or 'below' the level of the state. These are typically smaller-scale interventions, with action on energy supply issues aligning with the community's broader social goals around self-sufficiency and self-reliance (e.g. in the area of food production). At the micro-scale, communities like Brithdir Mawr in Wales (see **Example 2**) have a strong communal identity and shared values and its members have come together to become self-sustaining in energy alongside other material and spiritual needs (Figure 8.2). At a broader scale, the emergence in recent years of the Transition Towns Movement, a grassroots response to the dual challenges of anthropogenic climate change and energy scarcity, has begun to develop place-based energy initiatives that rely on locally available renewable energy sources (Transition Network 2014). Some of these initiatives go beyond switching to renewable sources to pursue 'energy descent' – i.e. the scaling down of net energy

---

EXAMPLE 2: SECURING ENERGY AT THE COMMUNITY SCALE

The community of Brithdir Mawr in Pembrokeshire, Wales, is an example of how a group of people have come together to live in a sustainable way, focusing on three components of sustainability: environment, people and education. The community is self-sustaining in terms of energy supply and demand, utilising a mixture of wind, photovoltaic and hydro-generation to meet their electricity demand. Solar hot water and wood chip-fired boilers provide hot water and heating, respectively. Brithdir Mawr is an example of a locally based, community-centred energy system that provides a high level of security from both grid-based price fluctuations and supply changes. Like other examples of its kind, it illustrates how community energy security can be achieved at the micro-scale. A major challenge is how to scale up such approaches so that they work in other spatial contexts and for a far greater number of people. See: www.brithdirmawr.co.uk/

Figure 8.2 The Brithdir Mawr community, Wales (photo credit: Brithdir Mawr community)

off-grid communities
a spatially defined group of individual households or a community with shared energy systems (for example, a district heating system) who are not connected to the national electricity or natural gas supply network, or grid. Individuals, households and communities deploy and develop strategies for their energy security, which may be low-tech in nature (for example, wood burning for home heating or building-in cold rooms for refrigeration) or utilise emergent technologies like solar photovoltaic panels and geo-thermal energy for electricity and heating, respectively.

use over time as part of a strategy of degrowth (**Chapter 6**), and in a manner that mirrors the ramping up of energy inputs since the Indus-trial Revolution (**Chapter 1**). Finally, communities and individual households who are off-grid (see **Case study**) have to ensure security of supply through providing for their own energy needs, whether through fossil fuels and/or renewable means (Figure 8.3). Here energy security strategies include the adoption of different lifstyle practices whose rhythms are in tune with the availability of given energy sources, such as wind, light, river flow or deliveries of oil and bottled gas.

In summary, communities have responded to energy security con-cerns in diverse ways, and through a number of different institutional and non-state solutions. Discourses of energy security are largely concerned with maintaining supply at currently experienced levels, so as to sustain existing energy services and 'acceptable' living standards. Only in a relatively small number of cases, and notably in off-grid communities, is this definition challenged, and alternative ways of living with radically lower inputs of energy put forward. For the time being, then, calls for energy security largely normalise and entrench existing social expectations about what living a comfortable, rewarding life means (Walker et al. 2007) and the ways in which new forms of energy production can be brought about (Foxon et al. 2005). In this way, energy security concerns often have the effect of locking-in demand for energy in the long-term, and in ways that perpetuate broader economic, political and environmental insecurities.

Living with energy insecurity

Fear of losing reliable and affordable energy supplies is at the heart of attempts to securitise energy. For many, however, energy insecurity is a structural condition, and one that can be made all the more difficult but the efforts of others to enhance their own energy security. It is often a result of the structural and social inequalities built into processes of decision-making that determine how and for whom resources should be distributed throughout society (Walker and Day 2012).

The phenomenon of energy insecurity may be observed at a range of scales. First, there are insecurities associated with political tensions between major economic blocks, such as that between the Russian Federation, former client states in Eastern Europe, and Western Europe (Klare 2002). The inter-meshing of politics, economics and inter-state energy insecurity is exemplified by the Russia–Ukraine gas crisis that commenced in 2006. Major energy disputes between Russia and Ukraine in 2006, and again in 2009, led to Russia's major gas supplier Gazprom shutting off supplies to Ukraine over an argument related to pricing (Bradshaw 2009). Much of the gas passing into Ukraine from Russia transits through the country en route to other European markets. During the first crisis (2006), Ukraine had diverted a large portion of the gas destined for other European markets to feed its own domestic demand, causing problems elsewhere in the continent. The crisis revealed a wider problem of energy reliance

Figure 8.3 Energy security via energy storage: a household wood pile

CASE STUDY: PAKS II – THE DYNAMICS OF ENERGY SECURITY AND SOCIETY IN HUNGARY

Contributed by Brian King, Central European University, Budapest, Hungary

Hungary sits in the centre of two overlapping regions with distinctive energy histories: Eastern Europe and Western Europe. This has created a unique regional context for energy security that intersects with domestic social and political issues within the country. Hungary was part of the communist bloc until the late 1980s and its economy and society were heavily influenced by ties to the Soviet Union. A legacy of that period is a post-Soviet energy land-scape characterised by import dependency, ageing infrastructure, and inefficient energy systems that facilitate energy poverty (Tirado Herrero and Ürge-Vorsat, 2012; Bradshaw 2013). In 2004, Hungary joined the European Union along with six other countries from the former communist bloc. Since that time, differences between Eastern and Western Europe – in terms of their energy histories, infrastructures and strategic objectives – have shaped the EU's internal and external energy policy (Austvik 2016). The EU has sought to develop a common energy policy for the 28 Member States, focused on energy market integration and decarbonisation, although the specific energy circumstances of individual member states mean there continue to be marked differences among national approaches to energy policy (Nosko and Mišík 2017).

Hungary's controversial recent decision to increase nuclear power capacity to meet its electricity needs provides a clear illustration of these systemic issues. The Paks Nuclear Power Plant, located in central Hungary, is the country's only nuclear station and provides around 40% of its electricity. It was constructed in the 1970s with a Soviet design and started operating in 1982. In 2009, two years before the Fukushima nuclear disaster in Japan, the Hungarian parliament approved an expansion of the Paks nuclear plant (Paks II) by an overwhelming majority. Energy security was a driving concern within Hungary and the wider region and had been heightened by the Russia–Ukraine gas crisis. Parliamentarians argued that increasing nuclear power generation was necessary to reduce the country's reliance on energy imports (Sarlós 2015). There was concerted public opposition to Paks II from critics of the ruling party, as well as from green organisations and energy policy think tanks. Scepticism intensified in 2014 when the contract for two reactor units went to Russian energy giant Rosatom, with a 10 billion euro loan from a state-owned Russian bank. Lack of transparency in the terms of the contract and financing details caught the attention of the EU and the wider international community, and led to the European Commission pursuing proceedings for infringement of EU law. Alongside accusations of corruption, opposition to the project pointed to deep flaws in estimates for costs and projected energy use put forwarded by the Hungarian government. Despite this, Paks II remained domestically attractive from the perspective of job creation and technological significance (Inne 2015).

The Paks II project embodies many of the tensions associated with energy security in Central and Eastern Europe, which centre on the competing interests of the EU, Russia, and national governments. The Rosatom deal exemplifies the ruling party's close relationship with Moscow and Hungary's broader 'turn to the east', which directly challenges EU goals of Member State cohesion (Isaacs and Molnar 2017). Apparent gains in energy security via increased domestic electricity production as a result of Paks II come bundled with a decades-long commitment to Russian-backed nuclear power. Investment in nuclear also stifles the adoption of renewable energy sources, and at a time when technological innovation and sustainable energy policies are sharply reducing the costs of electricity generation from renewable sources (Sáfián 2014). Debt from the expansion of Paks II creates further reliance on Russia, and could indirectly fall on the shoulders of Hungarian taxpayers through artificial increases in electricity rates imposed to recover costs. While Paks II is specific to Hungary, it exemplifies the energy security challenges faced by countries in Central and Eastern Europe, and how efforts to address these concerns drive future energy–society interactions within the region.

and insecurity amongst many European nations, which had been developing for some time. Although the dispute with Ukraine was ostensibly an economic argument over pricing, the incidents in 2006 and 2009 were both related to political tensions between Russia and Ukraine over the latter's closer ties with NATO and the EU, and Russia's conflict with Georgia. Russia's subsequent annexation of Crimea (from Ukraine) in 2014 highlights the level of tensions involved, and the profound consequences for energy insecurity not only in Ukraine but also much of Western Europe. Following these disputes, the EU has re-visited its energy policy which historically has relied heavily on Russian imports of natural gas. Some European countries have actively diversified their natural gas supplies, including greater use of imported LNG (see Figure 8.4).

Second, energy insecurity can refer to the energy vulnerabilities experienced by millions of individual energy users in both the global North and South, who have either intermittent supply or simply cannot afford rising costs of energy (Buzar 2007; Pachauri and Spreng 2011) (see **Chapter 7**). The challenge of energy insecurity at household and individual level can be conceptualised as a matter of **energy justice**, as it reflects important structural inequalities in society that influence access to adequate energy services. Research by Walker and Day (2012) shows how accessing adequate levels of energy services (e.g. for heat) can be highly problematic for low-income households in the UK, and in ways that mean poverty and energy insecurity reproduce each other. Those on low incomes often struggle to access the financial resources necessary to purchase energy services, reflecting the broadening gap between a small proportion of very highly paid individuals and those who earn at or just about the minimum wage, in a context where average wage increases cannot keep pace with fuel price rises. Furthermore, those least able to pay are often placed on the highest electricity and gas tariffs because of the unstable nature of their income, and consequent inability to access reduced tariffs that require proof of reliable bank funds. There are also inequalities in housing stock that mean some households live in poorly insulated properties with inefficient appliances,

energy justice refers to the ability of an energy system to distribute the benefits and costs of energy services in a fair manner, while offering representative and impartial decision-making. Energy justice research has explored the sites and locations in which injustices emerge within the energy chain (Bouzarovski and Simcock 2017), as well as the social groups that remain disadvantaged in this process. There is an emergent body of work on the legal and policy processes that allow for the recognition, remediation and alleviation of energy injustices (Walker and Day 2012).

Figure 8.4 The liquefied natural gas (LNG) carrier *Independence* arrives in the Baltic state of Lithuania (2014). The ship is a floating storage and regasification unit able to supply the Baltic states with LNG from multiple locations worldwide

which they are often unable to replace (see **Example 1**). As a result of these different processes, a large number of low-income households are unable to reliably access adequate levels of energy services.

CONCLUSION

In this chapter, we explored the concept of energy security and demonstrated the importance of asking for whom energy is made secure. Thinking about energy as a security concern may seem obvious: it reflects the way energy saturates social, economic and political life, and the long-distance supply chains that have developed for many of the energy carriers that power key sectors like transport and housing. However, the roots of the concept in the 1970s oil crises are historically very specific, and the shadow of these events continues to hang over contemporary understanding of energy security as a disruption to cross-border flows of oil and gas. Moreover, 'securitising' energy has profound effects: it is a powerful rhetorical device as it puts energy on the same level as a range of other threats to the security of the nation. By defending against disruptions to the status quo, it affirms prevailing norms of energy use in ways that can lock-in energy demand and ignore the broader (economic, political and environmental) failings of current energy systems. It is only once one steps outside the mainstream understanding of energy security that its material and scalar specificities become evident. Securing energy is a strategy undertaken by many actors (not only governments) and at scales above and below the nation state. These alternative actors and scales illustrate a wide range of strategic responses that go beyond ensuring the continuation of energy supply. They include approaches that seek to increase energy security by reducing energy use or adapting practices of consumption to match available supplies. Furthermore, when relieved of its association with cross-border flows of oil and gas, the concept of energy security can be used to show how insecurity with regard to energy is a structural condition for millions of people worldwide, and not only in the global South. Despite its limitations, then, energy security remains a useful term. We have shown how its definition has expanded over time in ways that subtly, yet significantly, change its focus. The shift towards acknowledging the affordability and acceptability of energy supply, in particular, increases attention to how and by whom energy is used, and who bears the social and environmental costs of making energy secure. The expanding meaning of energy security suggests the concept is flexible enough to accommodate important questions of vulnerability and justice, and that it can be applied to a broad range of actors and scales.

QUESTIONS FOR DISCUSSION

- What political and discursive strategies do nation states deploy to deliver energy security?

- Are there limitations to how energy security has been defined and interpreted by different groups in society?

- Why have alternative strategies for conceptualising and developing energy securities emerged in recent years and what are their characteristics?

- What are the characteristics of energy insecurity and at what scales does this operate?

- What broader social and economic changes are required to promote greater energy security for publics in the twenty-first century?

ACTIVITIES/POTENTIAL RESEARCH PROJECTS

- Consider what 'energy security' means for you. Make a list of the things that make energy a secure commodity – think, for example, about its price, availability, ease of use. What would make energy insecure for you?

- What are the ways in which individuals and communities develop understandings or energy security and pursue these by practical measures? What are the tensions and conflicts that emerge in this process for a particular community and how does this affect energy security for that community's future?

RECOMMENDED READING

■ Bridge, G. 2015. Energy (in)security: world-making in an age of scarcity. *The Geographical Journal* 181: 328–339.

This article highlights the historical conditions under which energy and security have become combined, the close association of energy security with crude oil imports, and the imprint that oil has left on the concept. It explores the securitisation of energy as an active project, highlighting ongoing work by social scientists to develop numerical and visual energy security indicators. The article concludes that it is important to understand what is at stake in the development of indicators of energy security, as they have the capacity to bring new geopolitical relations into being as much as to describe existing conditions.

■ Cherp, A. and J. Jewell. 2011. The three perspectives on energy security: intellectual history, disciplinary roots and the potential for integration. *Current Opinion on Environmental Sustainability* 3(4): 202–12.

The authors of this article argue that contemporary ideas about energy security have developed in response to several separate policy agendas. Behind these policy agendas they identify three distinct perspectives: a 'sovereignty' perspective on energy security with roots in political science; a 'robustness' perspective originating from engineering; and a 'resilience' perspective that draws on economics and complex systems analysis. The authors go on to suggest that any single perspective is inadequate, and that developing an interdisciplinary perspective requires all three be woven together.

■ Ciută, F. 2010. Conceptual notes on energy security: total or banal security? *Security Dialogue* 41: 123–144.

An early but influential conceptual contribution to understandings of energy security, this article draws on academic and policy-related sources to identify the existence of three logics of energy security: a logic of war, a logic of subsistence and a 'total' security logic. The author argues that the three paradigms are associated with different meanings, discourses and policy mechanisms.

■ Klare, M. 2002. *The resource wars*. London: Owl Books.

Klare adopts a geopolitical approach to understanding the relationship between conflict and resources, using energy and water as examples. Highlighting the global geo-political importance of resources, Klare's book starts by exploring the longstanding interest of the US in the Middle East and the importance of safeguarding its oil supplies. However, Klare is careful to highlight both the other energy resources that have been implication in conflict and the likely proliferation of such conflicts.

REFERENCES

Allen, J., W. Sheate and R. Diaz-Chavez. 2012. Community-based renewable energy in the Lake District National Park – local drivers, enablers, barriers and solutions. *Local Environment* 17: 261–280.

Austvik, O.G. 2016. The Energy Union and security-of-gas supply. *Energy Policy* 96: 372–382.

Bahgat, G. 2006. Europe's energy security: challenges and opportunities. *International Affairs* 82(5): 961–75.

Banister, D. 2008. The sustainable mobility paradigm. *Transport Policy* 15: 73–80.

Boyle, G., B. Everett and J. Ramage. 2003. *Energy systems and sustainability: power for a sustainable future.* Oxford: Oxford University Press.

Bouzarovski, S. and N. Simcock. 2017. Spatializing energy justice. *Energy Policy* 107: 640–648.

Bradshaw, M. 2009. The geopolitics of global energy security. *Geography Compass* 3: 1920–1937.

Bradshaw, M. 2013. *Global energy dilemmas.* London: Polity.

Bridge, G. 2015. Energy (in)security: world-making in an age of scarcity. *The Geographical Journal* 181: 328–339.

Buzar, S. 2007. The 'hidden' geographies of energy poverty in post-socialism: between institutions and households. *Geoforum* 38: 224–240.

Cherp, A. and J. Jewell. 2011. The three perspectives on energy security: intellectual history, disciplinary roots and the potential for integration. *Current Opinion on Environmental Sustainability* 3(4): 202–212.

Cherp, A. and J. Jewell. 2014. The concept of energy security: beyond the four As. *Energy Policy* 75: 415–421.

Chester, L. 2010. Conceptualising energy security and making explicit its polysemic nature. *Energy Policy* 38: 887–95.

Ciută, F. 2010. Conceptual notes on energy security: total or banal security? *Security Dialogue* 41: 123–144.

DECC. 2016. Annual fuel poverty statistics report, 2016 England. London: HM Government. Available online at www.gov.uk/government/uploads/system/uploads/attachment_data/file/637430/Annual_Fuel_Poverty_Statistics_Report_2016_-_revised_26.04.2017.pdf

Del Rio, P. and M. Gual. 2006. An integrated assessment of the feed-in tariff system in Spain. *Energy Policy* 35: 994–1012.

Department for Communities and Local Government (DCLG). 2013. *Feed-in tariffs: get money for generating your own electricity.* Available online at www.gov.uk/feed-in-tariffs.

Foxon, T., R. Gross, A. Chase, J. Howes, A. Arnall and D. Anderson. 2005. UK innovation systems for new and renewable energy technologies: drivers, barriers and systems failures. *Energy Policy* 33: 2123–2137.

Frontier Economics 2015. Energy efficiency: an infrastructure priority. Available online at www.frontier-economics.com/documents/2015/09/energy-efficiency-infrastructure-priority.pdf

Gale and Snowden Architects. 2014. *Oak Meadow: iconic, award-winning, affordable, low-energy homes in Devon.* Available online at www.ecodesign.co.uk/projects/residential/oak-meadow/

Gnansounou, E. 2008. Assessing the energy vulnerability: case of industrialised countries. *Energy Policy* 36: 3734–3744.

Gupta, Eshita. 2008. Oil Vulnerability index of oil-importing countries. *Energy Policy* 36(3): 1195–1211.

Hughes, Larry. 2009. The four 'R's of energy security. *Energy Policy* 37: 2459–2461.

Johansson, Bengt. 2013. A broadened typology on energy and security. *Energy* 53(May): 199–205.

Klare, M. 2002. *The resource wars.* London: Owl Books.

Kruyt, B, D. van Vuuren, H. de Vries and H. Groenenberg. 2013. Indicators for energy security. In *The Routledge handbook of energy security.* B Sovacool (ed.), London and New York: Routledge, pp. 291–312.

IEA. 2017. Energy security. Available online at www.iea.org/topics/energysecurity/

Innes, A., 2015. Hungary's illiberal democracy. *Current History* 114: 95.

Isaacs, R. and Molnar, A. 2017. Island in the neoliberal stream: energy security and soft re-nationalisation in Hungary. *Journal of Contemporary European Studies* 25: 107–126.

Mendoca, M. 2007. *Feed-in tariffs: accelerating the deployment of renewable energy.* London: Earthscan.

MetroSeville. 2014. MetroSeville website. Available online at www.metro-sevilla.es/en

Nosko, A. and Mišík, M., 2017. No united front: the political economy of energy in Central and Eastern Europe. In *Energy Union: Europe's new liberal mercantilism?.* S.S. Andersen, A. Goldthau and N. Sitter (eds), International Political Economy Series, London: Palgrave Macmillan UK, pp. 201–222.

Pachauri, S. and D. Spreng. 2011. Measuring and monitoring energy poverty. *Energy Policy* 39: 7497–7504.

Transition Network (2014) Transition Network website. Available online at www.transitionnetwork.org/

Sáfián, F. 2014. Modelling the Hungarian energy system – the first step towards sustainable energy planning. *Energy* 69: 58–66.

Sarlós, G., 2015. Risk perception and political alienism: political discourse on the future of nuclear energy in Hungary. *Central European Journal of Communication* 8: 93–111.

Sovacool, B. 2013. Introduction: defining, measuring and exploring energy security. In *The Routledge handbook of energy security.* B. Sovacool (ed.), Routledge, London and New York pp. 1–42.

Tirado Herrero, S. and D.Ürge-Vorsatz. 2012. Trapped in the heat: a post-communist type of fuel poverty. *Energy Policy*, Special Section: Fuel Poverty Comes of Age: Commemorating 21 Years of Research and Policy 49: 60–68.

Umbach, Frank. 2004. Global energy supply and geopolitical challenges. In *Asia and Europe – cooperating for energy security: a CAEC task force report.* François Godement, François Nicolas, Taizo Yakushiji and Institut François des Relations Internationales (IFRI) (eds), Paris: IFRI, pp. 137–168.

Walker, G. and R. Day. 2012. Fuel poverty as injustice: integrating distribution, recognition and procedure in the struggle for affordable warmth. *Energy Policy* 46: 69–75.

Walker, G., and P. Devine-Wright. 2008. Community renewable energy: what should it mean?. *Energy Policy* 36: 497–500.

Walker, G., S. Hunter, P. Devine-Wright, B. Evans and H. Fay. 2007. Harnessing community energies: explaining and evaluating community-based localism in renewable energy policy in the UK. *Global Environmental Politics* 7: 64–82.

PART THREE

Transitions, governance and futures

Part 3 centres on processes of change in energy systems. The insecurities, vulnerabilities and injustices outlined in Part 2 mean that energy system transformation is now a broad social objective and a specific policy goal for many states, cities, communities and firms. *Transition* and *governance* are open frameworks for understanding the socio-technical and socio-political changes necessary to this process of transformation; and *futures* is an invitation to think about the conditions of possibility for energy system change, and the actors, practices and imaginations associated with prospective energy pathways. Part 3 is simultaneously forward-looking and historical in scope: Past transitions (**Chapter 9**) draws on historic experiences of energy transitions at global, national, urban and household scales; and Future transitions (**Chapter 10**) considers the range of energy transition pathways now on offer, and the purposeful interventions and 'transition experiments' that attempt to create new energy futures. Issues of time, speed and space, and the plural and contested character of possible pathways, are central to the two chapters in Part 3. Both carry forward the critical social science perspective developed in Parts 1 and 2.

Transition describes a change over time from one state to another. It can be applied retrospectively, to name important structural changes that have occurred within an energy system; and it may be used prospectively, to identify desired trajectories of change. 'Transition' signals change in a particular direction but, on its own, says very little about the destination: it implies only that the end-state is different to what came before. This flexibility is part of transition's appeal as a concept but, because it is under-specified, first steps in understanding energy transition are to establish the direction of change and which components of the energy system it applies to. In the field of energy studies, the concept of transition typically refers to major shifts in the role of different fuels and conversion technologies within national energy systems and/or at the global scale. Classic examples include historic transitions from wood and water power to coal in many parts of Europe in the nineteenth century, and from coal to oil in the twentieth century. This focus on technologies and fuels alone, however, can "downplay the profound social and political disruptions such energy changes portend" (Laird 2013: 149). A socio-technical perspective, on the other hand, opens up the analytical imagination, embedding energy's key technological and material shifts in the nineteenth and twentieth centuries within broad processes of socio-economic and political change (such as industrialisation, urbanisation and the growth of consumer society). The chapters in Part 3 adopt this perspective, drawing on what is now a substantial socio-technical literature focused on the dynamics of energy transitions. This includes research on 'sustainability transitions' that combines concepts from technological innovation, industrial clustering and regional development pathways (e.g. Hansen and Coenen 2015); and theoretical frameworks such as the 'multi-level perspective' (MLP) on socio-technical transitions, that brings together ideas from evolutionary economics, science and technology studies, and institutional theory to explain how transitions occur (e.g. Geels 2005, 2014). An important insight from this work is that energy transitions require more than simply implementing the 'right' technology or switching fuels: the social, economic, and political arrangements within which new technologies and fuels emerge (and to which they subsequently give shape) are key to processes of transition.

Part 3 highlights the diverse and frequently divergent objectives to which the concept of energy transition is currently applied, including decarbonisation and the low-carbon economy; a future secured by fossil fuels; demand reduction; and expanding access to energy for development. By specifying the direction, dimensions and limits of transition, it is possible to distinguish between changes that appear substantial when judged by the level of investment in new technologies but which, nevertheless, 'go with the grain' by upholding existing social structures and practices; and those which may be smaller in scale but which involve (or enable) re-assembling social relations in significant ways. A distinction between 'conservative' and 'progressive' forms of transformation captures something of this dynamic, based on whether they close down or open up opportunities to challenge entrenched interests (Stirling 2014). Similarly, the notion of a 'just energy transition' speaks to the possibility that interventions to decarbonise the energy system can take a form that improves social justice, rather than affirming existing inequalities and distributions of social power (Newell and Mulvaney 2013).

Governance opens up questions of co-ordination and steering, and therefore complements Part 3's focus on the direction of energy transition. Its central concern is with how and by whom energy futures will be made. Governance is an explicitly broader term than government, and acknowledges how a wide range of non-state actors – private firms, social movements, non-governmental organisations, supra-national agencies, communities, consumers – influence processes of change within energy systems in significant ways. We refer to governance in this introduction to Part 3 to signal how chapters are interested in how change occurs in energy systems, the multiple transition pathways that are possible, and the open-ended character of energy futures. The term encompasses processes of innovation and experimentation 'from below', through which novel technical entities or social arrangements are tried out; and networks of action and influence that cut across different administrative and spatial scales, linking supranational organisations and transnational firms with national governments, cities and local communities. These 'multi-level' forms of governance have become particularly significant in the context of energy and climate change, and around global agendas like sustainable energy for all (Bulkeley and Betsill 2013). Governance's focus on the steering mechanisms for energy transition in the context of several different potential destinations also draws attention to processes of contestation and resistance, and the powerful role of incumbent actors within the energy system. In this way, Part 3 carries forward questions of justice and social power introduced earlier in the book.

Futures returns the book to the 'grand challenge' of energy system transformation outlined in the **Introduction**. We have chosen the plural form deliberately, to express the idea of multiple pathways leading on from the present, rather than a single solution or destination. The plural character of energy futures arises, in part, from the fact of spatial variation: geographical differences in how energy systems are organised, and whom they serve and effect, create the conditions in which a range of different energy futures may be imagined, pursued and achieved (Bridge 2018). A useful social science concept for thinking about these futures is the 'sociotechnical imaginary' (Jasanoff and Kim 2015). The term 'imaginary' here does not mean fantasy or fiction, but refers to shared conceptions of the world and the social meanings that attach to these conceptions. Energy systems are the subject of a wide range of 'sociotechnical imaginaries'. These, in different ways, relate specific technological assemblages (e.g. nuclear power, large-scale renewables, district heating systems, off-grid community solar) to future socio-economic and political possibilities, such as modernisation and development, green growth, energy poverty alleviation, and autonomy and self-governance. Calls for energy transition frequently revolve around distinctive sociotechnical imaginaries, and mobilise notions like urgency and feasibility to connect these imagined futures back to the present. The character of these different energy futures, and the pathways to them from the present, are a matter of broad social concern across a range of institutional and policy spaces.

INTRODUCTION TO THE CHAPTERS IN PART 3

Past transitions (Chapter 9) focuses on the mutability of energy systems over time, and how they are continually evolving and open to change. It examines significant transitions in the provision of heating, lighting, power and transport over several centuries. The chapter shows how these past energy transitions have rarely been a single event, but have occurred as a result of multiple and interlocking changes, many of them at different scales (global, national, urban, household). Energy transitions have also frequently involved creative disruptions in the realm of sensory experience (e.g. thermal comfort, lighting norms, speed of movement). Moreover, the spaces of work, leisure and pollution associated with new energy systems have contributed to the emergence of important cultural and political identities. The chapter explores how retrospective 'transition histories' can be an important source of information for understanding the prospects for and implications of contemporary energy transitions, such as pursuit of a low-carbon economy.

Future transitions (Chapter 10) asks the question 'What kind of energy transition and for whom?' The chapter reflects on how historic energy transitions have brought societies to the point where decarbonisation, energy security, universal access and energy justice are desirable goals. It highlights how energy transition is now being pursued as a political and economic project by states, cities, firms and local communities, and critically compares some of the different visions of transition within public and political discourse. The chapter considers how these multiple pathways are each shaped by dynamics of development, innovation and contestation, and how none on its own offers a perfect solution. Chapter 10's emphasis on multiple imperatives and pathways, and on the value of a critical social science understanding of the challenges and choices ahead, serves as a conclusion to the book.

REFERENCES

Bridge, G. 2018. The map is not the territory: a supportive yet critical reading of energy research's spatial turn. *Energy Research and Social Science* 36: 11–20.

Bulkeley, H. and M.M. Betsill., 2013. Revisiting the urban politics of climate change. *Environmental Politics* 22(1): 136–154.

Geels, F. 2014. Regime resistance against low-carbon transitions: introducing politics and power into the multi-level perspective. *Theory, Culture and Society* 31(5): 21–40.

Geels, F. 2005. *Technological transitions and system innovations: a co-evolutionary and socio-technical analysis.* Cheltenham: Edward Elgar.

Hansen, T. and L. Coenen. 2015. The geography of sustainability transitions: review, synthesis and reflections on an emergent research field. *Environmental Innovation and Societal Transitions* 17: 92–109.

Jasanoff, S. and S.H. Kim (eds). 2015. *Dreamscapes of modernity: sociotechnical imaginaries and the fabrication of power.* Chicago and London: University of Chicago Press.

Laird, F.N. 2013. Against transitions? Uncovering conflicts in changing energy systems. *Science as Culture* 22(2): 149–156.

Newell, P. and D. Mulvaney. 2013. The political economy of the just transition. *The Geographical Journal* 179(2): 132–140

Stirling, A. 2014. Transforming power: social science and the politics of energy choices. *Energy Research & Social Science* 1: 83–95.

Past transitions

9

LEARNING OUTCOMES

- Understand how energy systems have changed over time and identify key 'lessons' from past energy transitions.
- Consider energy transitions at global, national, urban and household scales and their implications for human experience.
- Highlight significant transitions in the provision of heating, lighting, power and transport over several centuries.
- Show how transition is a geographical process, involving the remaking of socio-spatial relations and the experience of space and time, and illustrate with examples.
- Encourage reflection on how the built environment provides a record of historic transitions.

Energy systems pose a paradox. Composed of seemingly durable infra-structures, technologies and resources, they are nonetheless subject to great change over time. In this chapter, we look backwards from the present to highlight significant ways in which energy systems have evolved. We show how these 'energy transitions' are wide-ranging, encompassing shifts not only in fuels and technologies of energy use, but also infrastructures, cultural practices, and the organisation of society more broadly. Past transitions have not been singular events, but socio-technical processes involving multiple changes in different sectors of the economy, at different scales and at different times. They reveal how, as today, earlier energy systems were shaped by concerns about energy access and environmental impact as well as technical debates over cost and efficiency. Retrospective 'transition histories' – including failed efforts to change the energy mix – can shed light on the scope and prospects of contem-porary energy transitions, such as the move towards a low-carbon society. An important lesson of past transitions is that energy transitions require more than simply implementing the 'right' technology or switching fuels (see **Chapter 10**).

Figure 9.1 Evidence of past energy transitions (clockwise from top left): 20th century coal mining waste in Nord Pas-de-Calais, France (photo credit: Jérémy-Günther-Heinz Jähnick); USS Texas, the last US battleship to be fitted with coal-fired boilers (photo credit: Rennett Stowe); Abandoned whale oil processing facility on South Georgia, Falkland Islands (photo credit: Liam Quinn); Corliss steam engine (1,400 horsepower) exhibited at the Philadelphia Centennial Exhibition in 1876 (credit: Benson John Lossing, Wikimedia Commons)

social metabolism
refers to the flows of materials and energy between nature and society. The term emphasises the (growing) scale at which materials and energy are appropriated from the environment, and the social and geographical patterns associated with their distribution and accumulation. The concept lies behind popular ideas like the 'ecological footprint' (of individuals, cities, firms or states), policy frameworks like industrial ecology and the circular economy, and critical approaches to uneven development at the world scale such as ecologically unequal exchange and ecological debt (see **Chapter 1**).

Past transitions demonstrate the significance of the social, economic, and political arrangements within which new technologies and fuels emerge, and to which they subsequently give shape.

Fortunately, time travel is not required to study past transitions. Evidence for them can frequently be found in the built environment, in the form of relic technologies of energy conversion, transmission and distribution that either cling on in use, or linger as defunct stocks of potentially recyclable materials (Wallsten 2015). Building designs and construction materials similarly betray historic practices of providing light, heat and power. In addition to inherited physical forms, a wide range of cultural practices, norms and aspirations (around personal mobility and thermal comfort, for example) carry historic energy transitions into the present. Together these landscapes and practices serve as testimony to the long-run **social metabolism** of energy and society and, with a little effort and a careful eye, may be 'read' as a partial record of prior energy transitions (Figure 9.1). In this chapter, we draw on a rich record of historical research to examine how earlier energy transitions have played out. The first half of the chapter explores 'supply-side' shifts in fuels and technologies of energy conversion.

In the second half, we focus on 'demand-side' changes in how energy services (heating, lighting, transport and power) have been experienced over time and reflect on their geographical and political implications.

HISTORIC ENERGY TRANSITIONS AT THE GLOBAL SCALE: THE GRAND FUEL SEQUENCE

We saw in **Chapter 1** how, at the global scale, the history of energy use over the past couple of centuries is marked by three distinctive trends. Energy availability has grown much faster than population; the energy intensity of economic output has fallen over time, as less energy is used per unit of GDP; and the rate at which energy can be put to work (i.e. power) has grown enormously, enabling large increases in the power that humans are able to command. Behind these three trends – and key to understanding them – are major shifts in fuel sources and the technologies of energy conversion through which societies put energy to work. Energy historians identify a **grand fuel sequence** (Figure 9.2) as societies initially shift their dependence from biomass to fossil energy sources and, subsequently, to reliance on a range of high quality energy carriers such as electricity and hydrocarbons (oil and gas). Overlaid on this sequence is an increasing reliance on inanimate (i.e. mechanised) means of converting energy into useful work, as machinery replaces animal and human muscle (Smil 2010a). Examples include the development of sails and water wheels, steam engines, electric motors, the

grand fuel sequence
a term popularised by Vaclav Smil (2010a) that describes the series of important shifts in a society's primary energy supply as it moves, first, away from a reliance on traditional biomass towards coal, and then from coal to a growing dependence on oil and gas. This generalised pattern applies to major industrial economies (UK, US, Germany, France, Russia, China, Japan, India) but is not, however, a universal experience: many Asian and Middle Eastern economies lacking coal have moved quickly from biomass direct to hydrocarbons, while large numbers of people in the global South continue to rely on traditional biofuels.

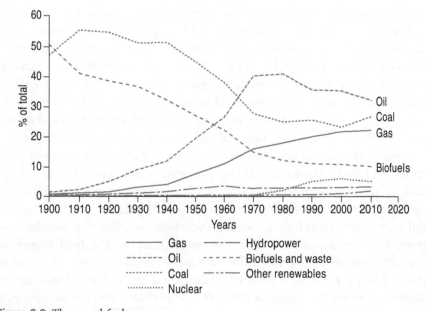

Figure 9.2 The grand fuel sequence

internal combustion engine
a machine in which a hot gas "is produced by combustion of fuel (petroleum or diesel) inside the engine" such as an automobile engine running on gasoline or diesel, or an aircraft jet engine (Smil 2010a: 56). By contrast, steam power is an external combustion machine "with water heated in a boiler and steam led into the piston chamber (or turbine)" (Smil 2010a: 56). An internal combustion engine is much lighter for a given output of power, reflecting the superior energy services that oil provides as a result of its liquid character and higher energy density. This enables higher speeds, and changes the economies of scale required for crossing space allowing the size of vehicle units to fall – from the train and tram to the automobile.

internal combustion engine and gas turbines, which over time have brought progressively larger outputs of power under human control. Importantly, these transitions in fuels and conversion devices have not occurred everywhere in the same way. Their uneven evolution over time has shaped contemporary inequalities in energy use at national, regional and global scales.

From an organic to mineral economy

Two major shifts transformed society's relationship to energy in ways that were globally significant prior to the mid-nineteenth century: the development and improvement of sedentary agriculture, and the tapping of fossil energy stocks (Wrigley 1990). Although very different in speed and form, both transitions made energy more abundant and increased available power, enabling societies to expand food surpluses, harness new landscapes and materials, and develop new knowledge and tools. While each occurred initially within a particular geographical context – domestication of plants in Southwest and Southeast Asia and Meso-America, and fossil fuel use in Europe – the step-changes in available energy and power associated with these transitions came to be felt much more widely.

At first glance, it may seem strange to regard the domestication of plants and animals as an energy transition. For pre-industrial societies, however, agriculture and the harvesting of wood for charcoal production were far and away the most important means for generating "the calories that drove work" (Kander et al. 2013: 8) The effectiveness of agriculture in capturing and transforming flows of solar energy was a primary limit on energy availability. Settled agriculture emerged around 6000 BC and its capacity to accumulate energy surpluses (in the form of food and fodder) marked a radical departure from hunter-gatherer societies (Simmons 2008). The development of farming practices and land management techniques (such as terracing and irrigation) accelerated the accumulation of useful biomass, substantially modifying vegetation cover and creating entirely new ecosystems such as cropland and upland pasture. The result was a "controlled solar energy system" that channelled a growing proportion of solar insolation into the socially useful forms of food, fodder and fuel (Krausmann 2013). The agricultural revolution in Europe pushed this process still further from the sixteenth century onwards, enhancing agricultural yields through the implementation of land use practices such as soil improvement (via liming, marling and manuring), drainage, and development of mixed pastoral and arable farms. Techniques of animal husbandry expanded available power, as did mechanical innovations (such as the horse gin) and improvements to farm implements and the design of harnesses for draught animals (allowing the coupling of multiple horses or oxen, for example). Biomass in the form of food, fodder and fuel provided around 95% of total energy in this 'advanced organic economy', with wind and water power making up the remainder (Wrigley 1990). Available energy and power expanded significantly, although ultimately both were constrained by the limits of human and animal muscle.

Society's turn to mineral energy stocks to supplement solar energy flows represents a decisive break with the organic economy (see **Chapter 1**). This transition occurred much more rapidly than the evolution of sedentary farming, taking a couple of centuries in most of Europe (slightly longer in Britain) and a matter of decades in other parts of the world. At the core of this transition is growing use of coal for heat and power. Coal eased two significant constraints of the advanced organic economy: the tight bond, in a society based on biomass, between energy availability and the land area dedicated to energy production; and an overwhelming reliance on muscle power (Wrigley 1990). Coal reduced the competition, inherent to the organic economy, between food supply and energy production. It also made energy plentiful and comparatively cheap by enabling the concentration of production and enhancing control over labour (Malm 2016). Freed from the land constraint, new industries (such as chemicals and steel) could develop that did not rely on the productivity of agriculture. The transition from biomass to coal was widespread but not everywhere was the same. In places with limited local coal resources, such as in Holland and Ireland, people turned initially to peat, and the transition to coal was delayed until the advent of bulk transportation via canal and rail.

Coal's decisive application, however, was to mechanisation via steam power. Steam massively increased available power, transforming possibilities in economies where the physical strength of workers and animals, supplemented by water and wind, had formerly set the limits of available power. The scale of change was enormous. Energy consumption in Europe, for example, increased fivefold on a per capita basis, following the shift from biomass to coal (Kander et al. 2013). It was estimated in 1880 that the capacity of steam power in France was equivalent to employing nearly 100 million additional workers. This represented around two-and-a-half 'energy slaves' for every person in France and, furthermore, coal and steam were "true slaves, the most sober, docile and tireless that could be imagined" (quoted in Wrigley 1990: 76). The effects of enhanced mechanical power were profound. Production could be concentrated in single buildings, driving economies of scale that magnified "many times over the quantity of work that was performed by each worker" so that "goods could be produced in such profusion that the conveniences of an earlier age came to be regarded as necessaries, and the luxuries no more than conveniences" (Wrigley 1990: 80). Steam power, based on the combustion of coal, proliferated across a wide range of industrial sectors to become what economic historians refer to as a **general purpose technology** with wide-ranging economic, geographical and socio-political effects.

general purpose technology
a technological development whose impact is felt economy-wide rather than on a specific sector, and which transforms productivity and the experience of space and time more generally. Classic examples are electricity and information technology. Steam power and electric motors can also be considered general purpose technologies.

Diversification and consolidation in the mineral economy: oil, electrification, and natural gas

A series of important transitions occurred within the fossil-fuelled energy economy, beginning in the second half of the nineteenth century (Figure 9.1). The overall effect of these supply-side shifts was to diversify energy resources and energy carriers, massively expand energy use, and consolidate the role of carbon-intensive forms of energy. Three energy transitions are particularly significant: the expanding role of oil in the twentieth century; the emergence of electricity as a brand-new energy carrier and its subsequent proliferation; and the post-war growth of natural gas's contribution to the global energy mix.

Oil has some pre-industrial uses, but it began to emerge as a modern energy source in the 1860s. As kerosene, oil substituted for the rendered bodies of lizards, fish and whales as a lamp fuel, and facilitated the mass-production of cheap paraffin candles. Crude oil subsequently made inroads into the power and transportation sectors where it competed with coal, electricity and biofuels (Bernton et al. 2010), and secured its contemporary dominance as a transportation fuel by the middle of the twentieth century. Oil was more flexible than coal: it could substitute for coal in many applications (as a fuel for raising steam in industrial or domestic boilers, for example) but it could also do things that coal could not. Oil's liquidity enabled easier handling and much greater precision in application: as a shipping fuel, for example, it avoided the need for an army of stokers, and the ability to flow oil precisely made possible internal modes of combustion (e.g. diesel or petrol engines). At the same time, oil's high-energy density delivered an equivalent energy content for about half the weight of coal (**Chapter 1**). As a result, oil opened up new markets for which coal was poorly suited, such as modes of transportation where issues of weight (of both fuel and the machinery for converting heat energy into motion) were significant like the petrol-driven car, motorcycle and aeroplane. However, oil was less ubiquitous in occurrence than coal, and the build-up of dedicated oil-consuming machinery required long-distance control structures to secure sources of supply. Until the end of the nineteenth century, the bulk of the world's oil came from a relatively small number of locations in the United States, Romania, Russia, Sumatra and Burma. A series of very large crude oil discoveries (in Mexico, Venezuela, the United States, the Persian Gulf, Canada and Soviet Union) in the first half of the twentieth century, and improvements in infrastructure (pipelines, shipping tankers, storage) significantly expanded and diversified supply. At the same time, low oil prices and improvements in refining techniques (e.g. catalytic cracking) expanded markets for new fuels and petroleum products (Bridge and Le Billon 2017).

Electricity emerged as a commercially viable energy carrier towards the end of the nineteenth century. The geographer and energy historian, Vaclav Smil (2005: 32) captures its significance:

> After millennia of dependence on the three basic energy conversions –
> burning of fuels . . . use of human and animal muscles, and the capture of

indirect solar flows in water and wind – large-scale generation of electricity introduced a new form of energy that has no rival in terms of its convenience and flexibility.

Many of the first electricity-generating stations were driven by water turbines, and provided electricity for illumination. Subsequently, a series of technological developments, notably AC current and voltage transformers, enabled long-distance transmission and encouraged the scaling up of electricity generation. The first commercial installation of an alternating current hydropower plant in the United States, for example, was the Redlands Power Plant in California (1893), which generated electricity from falling water using the highly efficient 'Pelton Wheel' water turbine technology. The close connection between electrical power and water continued in the first half of the twentieth century, with major hydropower schemes put in place in Norway (from the 1890s), the United States, Canada and the Soviet Union.

Electricity transformed established logics of industrial location based on coal and water power. It enabled industrial-scale power to be harnessed in locations with limited access to coal reserves, providing an alternative basis for industrialisation and regional development. Hydropower expanded rapidly in Norway, Sweden and Italy where it fuelled industrial development, and played a similar role in parts of Canada, Russia and the United States. Electricity was more geographically mobile than other energy carriers, with long-distance transmission making it possible to deliver power to remote or dispersed communities. In the first half of the twentieth century, rural electrification programmes in Europe, North America and Australia, for example, brought electrical light, heat and power to the countryside. Electrical power increasingly turned to steam turbines as global generating capacity expanded, as they offered much higher rotation speeds and maximum power capacities. Coal-fired power stations were initially sited close to markets (i.e. in cities) to minimise transmission losses, and operated as part of self-contained urban distribution systems. In the post-war period, however, large coal-burning electricity generating stations were increasingly constructed close to the source of coal to minimise transportation costs, and operated as part of regional or national transmission grids. These included 'mine-mouth' plants (such as in the Four Corners region of the United States) and tide-water locations based on the importation of coal by ship. Over time oil, nuclear fuel, gas and other locally significant sources (such as peat) contributed to electricity generation, although the basic tether binding electricity to heat production remained intact. More recently, and driven by climate policies targeting the carbon-intensity of electricity production, the growth of wind, hydro, and solar power capacity have begun to shift electricity generation away from its dependence on the production of heat.

Natural gas has historically been regarded as an unwanted by-product of oil production, and something to be disposed of at the site of extraction by venting or flaring. The availability of town gas (a by-product of coke production) limited demand for natural gas, and natural gas's widespread commercialisation required development of burner technology and construction of long-distance high-pressure pipelines to move it to urban areas (Smil 2010a: 37). Natural gas first became a significant part of the energy mix in the United States in the 1930s. Elsewhere, its uptake has been a feature of the second half of the twentieth century, associated with development of transmission infrastructure (pipelines, liquefaction and regasification terminals) and liberalisation of an energy policy that highlights the lower operating and capital costs of gas-fired power plants. Use of gas for electricity generation has soared, largely replacing oil (e.g. Japan in 1970s, Singapore 1990s) and increasingly displacing coal (e.g. in the UK from the 1990s, in the US in the 2010s). Climate policy has also played a role, as gas offers significant carbon-savings over coal and oil in electrical power generation (**Chapter 1**). However, the construction of new plant and equipment dedicated to gas combustion effectively 'locks-in' this fossil fuel for the next 20 to 30 years.

Seven lessons from past transitions

Several general conclusions can be drawn from retrospective histories of energy transition. First, energy transition is typically a slow process. Many historical shifts in energy conversion technologies and fuels have taken 50 years (two generations) or more to unfold, although rates of change have varied widely from place to place. The transition from biomass to coal took centuries in the British Isles, and was particularly slow in Ireland where peat continued to be the primary domestic fuel until the end of the nineteenth century. Most rural Irish households continued to use peat until the 1950s, when cheap oil and coal became widely available (Kennedy 2013). Yet the same transition, from wood to coal, took only decades in the United States (Fouquet and Pearson 2012). The 'grand fuel sequence' has also unfolded at various rates (and sometimes out of sequence) across different energy services, as we show in the second part of the chapter. In short, past energy transitions have rarely been a smoothly ordered sequence of changes. As Thomas Hughes, an influential historian of technological systems puts it, "invention, development, innovation, transfer, and growth, competition and consolidation can and do occur . . . but not necessarily in that order" (Hughes 1993: 57).

destabilisation
energy transition can be experienced by existing industries and incumbent firms (such as coal mining companies, oil producers and electrical utilities) as a destabilising political and economic process, characterised by declining financial resources and eroding public legitimacy.

Second, not everyone is a winner from transition. Energy transitions redistribute costs and benefits and so are not neutral in their economic and political effects. Assets and infrastructures associated with emergent fuels or conversion technology may appreciate in economic value in anticipation of expanding demand. Those tied to the incumbent, meanwhile, may do less well and experience devaluation and disinvestment so that, for firms and communities tied to incumbents, energy transition is experienced as growing uncertainty and **destabilisation**. Energy transitions also redistribute the environmental risks and social vulnerabilities associated with producing and consuming energy services. The transition to coal in the UK reduced the vulnerability of iron makers and householders to high wood prices, and delivered greater power into the hands of those who could afford it. But as coal increasingly took hold in the eighteenth and early nineteenth century, millions of men, women and children ended up working in appalling conditions underground. There were wider distributional effects too: "buildings disintegrated under the assault of sulphur dioxide emissions, slum dwellers suffered vitamin deficiencies from lack of sunlight, and deaths from lung diseases soared . . ." (Kander et al. 2013: 132). Newspapers in Manchester, England – where coal fuelled the boilers of textile mills, electricity-generating stations, railway locomotives, steam ships, offices and homes – commented on the inequalities at the heart of the transition to coal, warning that "pride in an Empire on which the sun never sets should be tempered by the reflection that there are courts and slums at home on which the sun never rises" (quoted in Kander et al. 2013: 143). Distributional effects like these have frequently become the object of political struggle, from the labour movement in the early twentieth century (in which coal miners and other 'energy workers' in steel, shipping and railways played a pivotal role), to clean air campaigns and the wider environmental movement in the post-war period. In short, the direction, speed and effects of past energy transitions are negotiated outcomes, a product of socio-political struggles over how the 'goods' and 'bads' of energy service provision should be distributed over space and time.

> Described by Turnheim and Geels (2012) as the 'flipside' of energy transition, destabilisation provides a novel perspective. For example, it highlights how an effective response to climate change requires not only nurturing low-carbon alternatives but also deliberate interventions aimed at weakening society's cultural, political and economic commitments to fossil-fuel industries. It also suggests incumbent firms will initially resist such efforts at destabilisation.

Third, incumbents fight back. Those on the 'losing' side of an energy transition – for example, businesses heavily invested in canals at the start of the railway era, or holding major coal assets as heat, mobility and power generation shifted towards liquid fuels – are unlikely to cede ground without a fight. Technological responses include 'last-gasp' effects, as competition from new sources accelerates innovation among incumbents. Classic examples include the development of iron-hulled clipper ships with steel masts and multiple sails

---

EXAMPLE 1: THE MULTI-LEVEL PERSPECTIVE ON ENERGY TRANSITION

The multilevel perspective (MLP) is a social science research framework for understanding large-scale, socio-technical changes in energy, food and transport systems over time. It was developed initially in Europe in the mid-2000s, closely associated with the work of the Dutch-trained scholar Frank Geels (see Geels 2005). The MLP examines transition as an outcome of interactions over time among three different scales of organization: niche-level innovations, established regimes and an exogenous landscape. Recent work with the MLP seeks to infuse its account of transition with a richer political-economic and spatial analysis (see Baker et al. 2014; Gailing and Moss 2016).

reverse salient
a term borrowed from
military strategy, where it
refers to a backward bulge
in an advancing front line,
by the historian of technology
Thomas Hughes (1983). It
describes the technical
component and social barriers
which hold back the evolution
of a technological system.

path creation
a concept from innovation
studies describing the
political, economic and
cultural processes through
which new technological
pathways emerge. In
reference to energy systems,
it emphasises the possibility
of creating alternative
pathways to those occupied
by dominant energy resources
or technologies. By enabling
alternatives to emerge and
supporting their proliferation,
path creation can be an
important process in energy
transition. Contemporary
policy efforts to support
renewable and low-carbon
forms of power and heat are
examples of path creation,
as are the historical
promotion of nuclear power
and efforts throughout the
twentieth century to develop
synthetic liquid fuels from
coal and biomass to replace
the need for oil (Johnson
et al. 2016). See also path
dependency (Chapter 3).

in the second half of the nineteenth century after the onset of marine steam power (Geels 2005); and the contemporary promotion of 'clean coal' in the context of climate change and rapid uptake of natural gas and renewables (Markusson et al. 2017). Political responses include defensive efforts to limit or resist further systemic change, via lobbying for policies that protect the interests of incumbents rather than supporting emergent rivals. In the late nineteenth century, railway companies sought to stifle competition from road haulage by supporting vehicle speed restrictions and undermining calls for improvements to road surfaces (Fouquet 2008). An important lesson, then, is that "transitions are just as much about the decline of incumbent industries, as about the rise of new ones" (Fouquet and Pearson 2012: 5). Hughes (1993: 73) concept of the **reverse salient** is helpful here, as it describes components of a system that "have fallen behind or are out of phase with the others" and which, as a consequence, exert a drag on the system's evolution.

Fourth, growing energy efficiency has tended to expand energy consumption rather than reduce it. Over time, technological change has enabled societies to wring more useful heat, light or power from each unit of energy consumed. Energy services have become cheaper over time, as more can be obtained from an equivalent amount of fuel or less fuel used for a fixed amount of energy service. Overall energy consumption has surged in response. In their work on the UK, for example, Fouquet and Pearson (2006) found that the cost of artificial illumination fell 3,000-fold between 1800 and 2000. The consumption of lighting services expanded 40,000-fold as a result, a phenomenon known as the 'rebound effect' or Jevons paradox (see **Chapter 6**).

Fifth, energy transitions tend to be evolutionary rather than revolutionary (Melosi 1982). Past energy transitions have been characterised not by rapid or total substitution, but by diversification, slow displacement and growing spatial complexity. The dramatic growth of coal consumption in Europe during the nineteenth century, from less than 30% to over 80% of total energy supply, took place against the backdrop of a growing horse population, expanding fodder crop production, and growing investment in wind and water power (Kander et al. 2013). Shifts from one fuel source to another have also required integrating different geographies into energy supply chains and networks. Early biomass systems, for example, had relatively short supply chains, although growth pressures saw these extend over longer distances: the English iron smelting sector, which used charcoal derived from wood as its fuel source, imported wood from the Baltic Sea in the Middle Ages to ease local supply constraints. Coal supply chains were relatively short on the coalfields themselves, but the broader transition to coal in the nineteenth century rested on expanding the fuel's reach beyond the coalfield via the development of canals, railways

CASE STUDY: NUCLEAR POWER – YESTERDAY'S FUEL OF THE FUTURE?

History offers several examples of societies consciously trying to change the prevailing fuel mix to meet broad social objectives. Some of these efforts to achieve energy transition have been strikingly successful, such as the shift from town gas to natural gas in many parts of Europe in the 1960s, or programmes of rural electrification from the 1930s onwards. Other attempts to engineer broad shifts in energy use have been less successful. These 'failed transitions' can provide useful lessons in the context of efforts to meet decarbonisation objectives and transition to a low-carbon economy. The chequered history of nuclear power is a case in point: once embraced as the fuel of the future, many countries are now actively phasing out nuclear power or scaling back plans as the economics of investment are unattractive.

In the immediate post-war period, a handful of industrial states sought to develop nuclear power. The technical success of nuclear fission, and the prospect of almost limitless cheap electricity, led many more states to look to nuclear power as a way to meet growing demand for electricity (Figure 9.3). The proliferation of commercial nuclear reactors was facilitated by the US 'Atoms for Peace' programme, announced by President Eisenhower at the United Nations in 1953, by which the United States sought to contain the destructive potential of atomic energy by managing the transfer of reactor technology to other countries (Jasanoff and Kim 2009); and by a similar programme of technology sharing by the Soviet Union within countries of the Committee for Mutual Economic Assistance (COMECON). Via technological, financial and political support, these initiatives at path creation sought to bring about an expanding civilian use of nuclear power. At the same time, several states embraced nuclear reactor technology as part of a strategy of techno-scientific modernisation and national development: South Korea, for example, aligned nuclear power with its national goal of rapid techno-economic transformation, and France's development of nuclear power technology from the 1950s has to be understood in the context of the country's search for a new national identity in the post-war period (Hecht 1998).

The 'oil shocks' of the 1970s further consolidated interest in nuclear power, by elevating energy security as a policy concern (see **Chapter 4**). Many oil importing states initiated programmes of energy transition to decrease reliance on imported oil, in response to the decision by the Organisation of Arab Petroleum Exporting Countries to impose an embargo on crude oil shipments in 1973 (in retaliation for US support of Israel during the Yom Kippur War). Accelerating the development of nuclear power was a common theme within these energy plans. In the United States, for example, President Nixon announced in 1973 a goal of achieving energy self-sufficiency by 1980 (Holl 1982). 'Project Independence' included a rapid expansion of nuclear power, along with investment in coal-to-oil technology, energy conservation initiatives, and expediting the Trans-Alaska oil pipeline. In total, 200 new nuclear power plants, 150 coal-fired power stations, 250 new coal mines and 20 synthetic fuel plants were planned to be built in 10 years (Bordoff 2017). The goal of 'energy independence' has resonated within US policy circles ever since, but the imagined scale of nuclear development did not materialise. Many plants were cancelled during the 1970s and 1980s and, until construction began at the Virgil C. Summer power station in South Carolina

A pound of fuel to light Chicago

THAT'S ALL THE URANIUM needed to produce atomic power equal to the energy in 3 million pounds of coal. It could light the city of Chicago for a full day!

Atomic research is focused on developing an economical way to produce electricity from atomic energy. Scientists at Oak Ridge National Laboratory, which Union Carbide Nuclear Company operates for the Atomic Energy Commission, have already built experimental power producing reactors that are serving as a guide to commercial atom power plants.

Peaceful uses for the atom have also been found in the diagnosis and treatment of disease. Radioactivity is uncovering important facts about plant and animal growth. Industry uses the atom's radiation to control production processes, to test product quality, and for research.

The challenging field of atomic energy is not new to the people of Union Carbide. They have been pioneering in every phase of this exciting business — from the mining of uranium ore to harnessing the atom for our future comfort and well-being.

FREE: *To learn more about the atom and the tremendous strides made in the peaceful applications of atomic energy, write for the illustrated booklet "The Atom In Our Hands."*

UNION CARBIDE
AND CARBON CORPORATION
30 EAST 42ND STREET   UCC   NEW YORK 17, N.Y
In Canada: UNION CARBIDE CANADA LIMITED, Toronto

──────────── *UCC's Trade-marked Products include* ────────────

ELECTROMET Alloys and Metals	NATIONAL Carbons	ACHESON Electrodes	SYNTHETIC ORGANIC CHEMICALS
HAYNES STELLITE Alloys	PREST-O-LITE Acetylene	PYROFAX Gas PRESTONE Anti-Freeze	UNION CARBIDE Silicones
UNION CARBIDE	EVEREADY Flashlights & Batteries LINDE Oxygen	BAKELITE, VINYLITE, and KRENE Plastics	Dynel Textile Fibers

Figure 9.3 The promises of nuclear fission: 'A pound of fuel to light Chicago' (1956) (Courtesy of National Atomic Museum Foundation)

in 2013, no new nuclear reactors had been constructed in the US since the late 1970s. Accidents at Three Mile Island (United States, 1979), Chernobyl (Ukraine, 1986) and Fukushima (Japan, 2011) shifted public opinion in many countries, while the long-term problem of nuclear waste disposal, rising costs and limited profitability relative to other forms of power generation has deterred investment.

More recently, growing demand for electricity and the potential of nuclear power as a low-carbon source of electricity led to anticipation of a 'nuclear renaissance'. However, this has failed to materialise at scale. In the United States, for example, the low price of natural gas and the inability of nuclear power to attract investment have limited the development of new nuclear power capacity: construction of the Summer power station in South Carolina, initiated in 2013, was abandoned in 2017. Other countries have actively decided to phase out nuclear sources of electricity generation. Germany's experience, from a rapid build-up of its nuclear fleet in the 1960s to complete phase out by 2022, highlights this pattern of policy-driven phase out. Germany has closed half of its nuclear stations following the partial meltdown at Fukushima (caused by an earthquake and subsequent flooding of the reactor site), while Belgium, Spain and Switzerland also plan to phase out their nuclear power stations. In the wake of Fukushima, opposition to nuclear power has grown in South Korea and Japan. China initially imposed a safety review and moratorium on new plants, although it had lifted this by the end of 2012 and scaled up plans for the contribution of nuclear and renewables to the electricity mix in an effort to address air pollution from coal-fired power plants and reduce reliance on oil imports. France, which generates most of its electrical power from nuclear power and has the largest fleet of reactors outside of the US, is also seeking to reduce nuclear power's role in electricity generation in favour of renewables: a law adopted in 2015 aims to reduce nuclear power from 75% of electricity generation to 50% by 2025, and the French nuclear industry is bracing itself for a future focused on decommissioning and waste management rather than new construction.

and bulk marine shipping. Electrification, and the growth of town gas, in the nineteenth century coupled long-distance coal supply chains to local systems of energy production and distribution. Rapid growth in urban energy consumption, however, meant that by the mid-twentieth century these urban-based systems faced constraints and they became integrated within (inter)national gas and electricity grids (Rutter and Kierstead 2012).

Sixth, energy consumers like a bargain. Past energy transitions have been driven, on the whole, by consumers searching for sources of energy and power that confer economic advantage. New energy conversion technologies and fuels have been taken up in the past because those paying for energy services have accrued direct benefits from switching. Sometimes, this has taken the form of lower costs, although often the decision to switch also includes other 'benefits' such as higher quality energy services and/or greater convenience (e.g. more easily controlled, or less time involved in gathering and preparing fuels and/or cleaning up after their use). A low-carbon energy transition is distinctive in this historical context. The primary benefit of shifting towards lower-carbon sources – mitigating the effects of climate change – will be felt at the level of society as a whole rather than individual energy users. As we point out in **Chapter 10**, it is therefore a 'willed' form of transition that requires careful design and implementation of policy to achieve.

Finally, the future of energy has a fascinating history. Cultural historians highlight the enduring social power of energy transition as a utopian promise. George Basalla (1982: 27) pointed out some time ago how this promise of transition takes on a standard form because, at first, "any newly discovered source of energy is assumed to be without faults, infinitely abundant, and to have the potential to affect utopian changes in society". This was initially the case with coal, with writers and artists in early nineteenth century Europe both thrilled and repelled by the socio-spatial transformations that coal and steam power were beginning to unleash. As the negative effects of coal became increasingly apparent in the late nineteenth century, however, hopes that once had revolved around the transition to coal became displaced onto hydropower. As coal before it, hydropower was extolled in the early twentieth century. It was a super-abundant, low-cost and life-enhancing (less polluting, labour saving) 'white coal' able to replace the drudgery, toil and health problems associated with mining and urban air pollution. Whether it is coal in the mid-nineteenth century, hydropower in the early twentieth century, nuclear fission in the post-war period (see **Case study**), or nuclear fusion, hydrogen technology and bioenergy today, energy transition has always been the 'next big thing' capable of ushering in widespread technological and social change.

EXPERIENCING TRANSITION: KEY SHIFTS IN DEMAND FOR ENERGY SERVICES

This section considers how the population at large has experienced past transitions, by focusing on four fundamental end-uses that drive energy demand: light, heat, power and mobility (Fouquet 2008). We show how the economic and social significance of these energy services, and the combination of appliances and fuels through which they are provided, has changed dramatically in the last couple of hundred years. We draw on national level data (Fouquet 2010) supplemented by reflections on the urban and household scales. Because of their direct connection to personal and family histories, the latter are particularly useful for understanding the scale and pace of change. An average European household today, for example, uses around 200 times as much light each year than in 1800, a far greater increase than in food or clothing (Fouquet 2008). At the same time, and echoing the paradox with which this chapter opened, the energy systems providing this massively expanded level of energy services have grown increasingly invisible so that "touching, feeling, seeing, or smelling an energy source is the exception rather than the rule" within the global North (Hirsh and Jones 2014: 109).

Lighting

Light is integral to domestic life, industrial production and commercial spectacle. Illumination has been central to safety and security across time, from repelling wild animals, to coastal lighthouses and urban street lighting. Highly controllable forms of light are now part of the circulatory system for the information economy, via laser scanning and fibre optic technologies that code and transmit data as optical pulses. For most of human history, the availability of light has been governed by the rise and fall of the sun, supplemented by moonlight and firelight. Over the past couple of centuries, however, both the desire for illumination and the ability to provide lighting services on demand have been revolutionised. Lighting has become cheaper, more abundant and infinitely more controllable in delivery and, as a consequence, its applications have proliferated with profound economic, cultural and environmental consequences. Fouquet (2010) identifies three significant transitions in lighting after 1800, based on the historic experience of the UK: a shift from candles and oil lamps to gas; from candles to kerosene lamps; and from gas to electricity (Figure 9.4). Their combined effects were to dramatically drive down the cost of lighting, widening access to lighting services and encouraging an expansion in lighting's applications, from public illumination and commercial businesses to the private home. Transitions were also associated with an improvement in the quality of lighting services (brighter, more reliable and controllable, and with fewer unwanted effects like soot or odour), and a social and spatial displacement of the work involved in providing lighting services. Lighting, and particularly the rapid proliferation of electric lighting during the early twentieth century, also created new perceptions of time, affecting both economic production cycles and cultural life-styles.

Candles and oil lamps were the primary source of artificial illumination before the mid-nineteenth century. Initially, candles were made from animal fats (tallow, beeswax) or vegetable oils. Oil lamps were adopted in wealthier households and commercial establishments, and were used to provide public lighting services such as lighthouses and urban streets. These burned fish oil, whale oil or a vegetable oil like colza made from rapeseed, cabbage or kale. The introduction of mass-produced, paraffin candles (made from petroleum) in the second half of the nineteenth century drove down costs of illumination and ensured candles remained a key source of illumination for many, notwithstanding the development of oil lamp and gas lighting technology. The advent of 'gaslight' at the beginning of the nineteenth century stemmed from efforts to find a market for gas released during the process of making coke for iron manufacturing. This was not so-called 'natural gas' drawn from underground wells but 'town gas' made from the refining of coal. Town gas was substantially cheaper than candles, but came with large upfront costs for laying distribution pipes and fitting gas burners, which initially restricted its use to public or commercial lighting. From the eighteenth century onwards, Town Improvement Acts increasingly required the provision of lighting in public

243

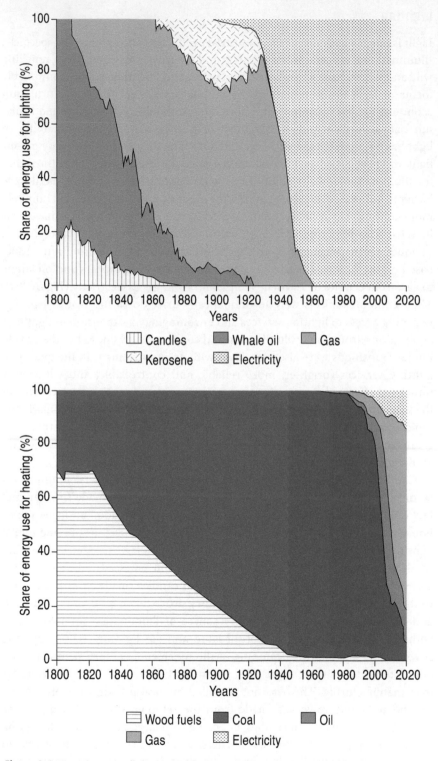

Figure 9.4 Transitions in lighting and heating services, UK 1800–2020 (based on Fouquet 2010)

places. Public illumination was initially provided by oil lamps, but gas light was demonstrated in Paris in 1801 and installed in Baltimore in 1816, the first city in the United States to have gas streetlights. Gas lighting was the first 'net-worked' provision of urban energy services, requiring development of fixed infrastructure linking multiple parts of the city, centralised production and distribution of gas, and standardised appliances and contracts. It displaced much of the work of securing energy services away from the household as, unlike lamps or candles, gas required no manual re-stocking. By 1849, gas accounted for over three quarters of all illumination in Britain, with 700 gas works providing lighting services for public streets, commercial buildings and wealthier homes (Fouquet 2008).

Networked gas increasingly replaced the decentralised and labour-intensive process of candle and rush-light illumination. By the end of the nineteenth century, urban gas companies had extended their networks and consolidated the role of gas in urban lighting (Fouquet 2010). Candles, which had provided over 90% of all illumination in 1800, shrank to less than 1% of all lighting services, although candle use expanded overall, providing around 50% more light than at the beginning of the century. The quality of gas light improved over time, particularly when emerging competition from electrical lighting drove development of the incandescent mantle in 1885, which fitted around the burner and produced a bright white light when heated. Where town gas networks were not developed – and among poorer people who could not afford gas – candles were replaced by kerosene lamps, which were cost-competitive with candles but provided superior illumination. Both these transitions – from candles to town gas (derived from coal), and from candles to kerosene (derived from crude oil) – demonstrate the shift from an organic to a mineral-based energy system in the area of lighting, with associated improvements in the cost and quality of energy service provided. The shift to electricity for lighting was gradual, and gas light remained a feature of some homes (in the UK) as late as the 1950s. Like town gas, electricity was pioneered for public and commercial applications. It was initially associated with arc lighting, where a spark passing between two electrodes provides illumination, and first used for public display in 1878 as part of the *Exposition Universelle* in Paris. Arc lighting produces an intensely bright light well suited to outdoor illumination but that was hard to control and not suitable for indoor use. Down-scaling illumination so it could be used indoors became possible with development of the incandescent light bulb, in which light is provided by heating a wire in the presence of a gas. Nonetheless, electric lighting was relatively slow to take root in the domestic home: only 6% of UK households had an electricity connection in 1919, for example (Fouquet 2008). Completion of a national electrical grid (1933) and a growing availability of electrically powered appliances spurred growth in electrical lighting so that, from the mid-1930s onwards, electricity was the dominant means of lighting. Today, an economy like the UK consumes around 25,000 times as much light overall as in 1800 (Fouquet 2008: 216). Contemporary developments in lighting technology, such as compact fluorescent lighting and solid state devices such as light emitting diodes (LED), create new opportunities for lighting applications, while also consolidating expectations of illumination (at work, home and in public space) that evolved during the nineteenth and twentieth centuries, and are now tightly bound to norms of public safety, commercial possibility and aspirations for the good life.

Heating

Heat is one of society's primary energy needs. In the home, heating provides thermal comfort and is used to cook food and warm water. Industrial and commercial applications of heat are many and various, and the capacity to manipulate and precisely control temperature has been central to the development of new industrial processes, from brick and glass-making to food processing and specialised steels. The 'grand transition' from an organic to mineral economy occurred first in heating, rather than lighting, transport or power. Coal was increasingly substituted for wood for home heating during the sixteenth century in Britain, due to a combination of rising woodfuel costs and falling prices for coal. Two obstacles to the use of coal in heating, however, were its effects on the material being heated and the problem of coal smoke. Domestic heating and cooking using coal required major changes to house design (see **Example 2**), while its application to a wide range of industrial processes took time. In iron production, it took over 200 years for a mineral replacement (coal, in the form of coke) to replace charcoal derived from wood (Fouquet 2008). The transition from wood to coal in heating started early, therefore, but took a long time to play out (Figure 9.4).

EXAMPLE 2: INVENTING THE COAL-BURNING HOUSE

Coal has a history before its association with the Industrial Revolution. In 1700 around half of all the coal burned in Britain was used to heat people's houses. Coal had already begun to replace wood by 1550, but its proliferation as a home heating fuel in the sixteenth and seventeenth centuries was not straightforward. Allen (2012: 20) shows how "switching fuels (from wood to coal) created complex design problems . . . which began with the design of the house'." Medieval houses typically burned wood in a low, open hearth located in the centre of a large room or hall, with smoke rising to the rafters and exiting through a hole in the roof. Coal, on the other hand, required a small and enclosed hearth with a raised grate to burn effectively. A tall and tapering chimney was also needed to induce a draft and remove sulphurous fumes from the coal smoke. The end result was a "house designed around a central chimney with back to back fireplaces on the ground and first floors. (It) could burn coal and warm the house without filling it with smoke" (Allen 2013: 13). Although the coal-burning house was invented in sixteenth century, the design needs of coal for domestic heating continued to influence the physical structure, layout and appearance of British housing stock well into the twentieth century (see Figure 9.4).

Coal was the primary heat source for cooking for most of the nineteenth century. Gas stoves were developed from the 1870s onwards, as gas companies started to seek alternatives to the lighting market in which electricity was starting to erode the dominance of gas. Gas was promoted as a 'cleaner' domestic fuel to coal in the context of growing concern with urban air pollution. The development and mass marketing of gas stoves, which were offered to households by gas companies on a rental basis, ensured the rapid expansion of gas cooking in the early twentieth century: in Britain, three-quarters of homes had gas stoves by 1939, whereas the same proportion had relied on coal only 40 years before (Fouquet 2008). Coal held on much more forcefully in the space-heating market than it did in cooking, however, although increasingly households turned to coke rather than raw coal, which was cheaper and, at the household scale, burned with less smoke. The introduction of clean air legislation (e.g. the UK Clean Air Act in 1956) consolidated this shift and "finally excluded the use of raw bituminous coal in homes, after about five hundred years" (Fouquet 2008: 85).

Since the middle of the twentieth century, demand for household space heating has been increasingly met via central heating technology, which involves a distributed set of heat delivery devices (radiators, air vents) throughout the house connected to and controlled via a single heat source (e.g. a boiler or burner unit), with temperature maintained within a relatively narrow range. Central heating technology can be supplied by a range of different energy sources, of which natural gas, oil, and electricity are the most common. In the post-war period, households in North American and Europe increasingly turned to natural gas as a source of central heat providing the "first major switch of heating fuel in three hundred years" (Fouquet 2008: 87). District heating performed effectively performed the same switch, scaling up central heating beyond the individual dwelling unit to encompass multiple homes within an urban neighbourhood. Central heating technology has significantly expanded the "total volume of deliberately conditioned space . . . around the world (that uses) energy to keep indoor air at a steady 'room temperature', whatever the weather outside" (Shove et al. 2014: 116). Together with the development of mechanical cooling and air-conditioning, it has changed the way we think about thermal comfort, driven changes in household building design, and facilitated new patterns of indoor social behaviour. In the early nineteenth century, 15°C was considered an ideal temperature conducive to good health and, during cold periods, occupants of a household donned thick clothing and clustered in the kitchen around the main source of heat (Fouquet 2008). Today, indoor temperature norms are in the low 20s, and central heating has enabled more dispersed patterns of room occupancy around multiple heat sources (including, for example, the rise of the bedroom as a space of independent relaxation rather than solely sleeping. Air-conditioning is regularly built into new commercial and institutional buildings (e.g. hospitals and schools), even in temperate climates, "as a means of delivering and ensuring operating conditions required by an increasingly wide range of heat-sensitive technologies" (Shove et al. 2014: 118).

Power

Pre-industrial demands for power were closely linked to the agricultural economy. Most was supplied by human and animal muscle, supplemented by wind and water (Figure 9.5). The substitution of horses for oxen in northern Europe increased available power by about 50%, while development of the horseshoe and improvements in harness design further expanded available power (Fouquet 2008). Wind and water mills mechanised processes for treating harvested crops (e.g. grain mills) and treating wool and other fibres for textile manufacturing (e.g. fulling mills). Coal's application to power became a defining feature of the nineteenth century but was slow to develop. Newcomen pioneered the 'atmospheric engine' in 1712, the first to harness heat from burning coal to mechanically pump water. By the 1760s, these stationary engines provided about half of all mechanical power at British coal mines, supplementing water power and animal and human muscle. Development of the condensing steam engine by James Watt (1768) drove a step-change in efficiency: unlike Newcomen's engine, Watt's design avoided repeatedly cooling and reheating the cylinder to drive the piston and used much less fuel. It became cost effective to drain deep mines, and to operate steam engines at locations where fuel had to be brought in from some distance. Subsequent innovations transformed the steam engine's reciprocal (i.e. up-down) motion into rotation, making it possible to drive wheels (winding gear, flywheels and cams) rather than solely pumping. Steam gradually replaced water as a source of power in iron production (for bellows and hammering), coal and metal mining (lifting, crushing), and in textile and paper manufacturing. Steam power enabled the geographically uneven accumulation of a large stock of fixed capital, in the form of the massive textile and steel mills, deep mines and urban infrastructure built in this period (Kander et al. 2013).

Although steam power became dominant in the nineteenth century, it never replaced wind and water. Development of the water turbine after 1830 squeezed much greater power from falling water than had been possible via water wheels and, by the middle of the nineteenth century, the largest water turbines were as powerful as those driven by steam. Towards the end of the nineteenth century, however, steam power attained higher efficiencies and larger power outputs by switching from piston engines to steam turbines. Steam turbines were quickly taken up in the generation of electrical power, where they continue to have key applications regardless of whether the primary energy source coal, gas, oil or nuclear fission. They also found widespread application in the maritime sector where, in the early twentieth century, they competed against both steam power and emerging marine diesel engines. There were similar (although commercially unsuccessful) experiments using both steam and gas turbine engines in rail transport, particularly in countries like Sweden which had limited domestic coal reserves. Steam turbines are now among the world's most powerful machines. The largest are able to generate more than a gigawatt of power, a millionfold increase on the earliest steam engines (Smil 2010a).

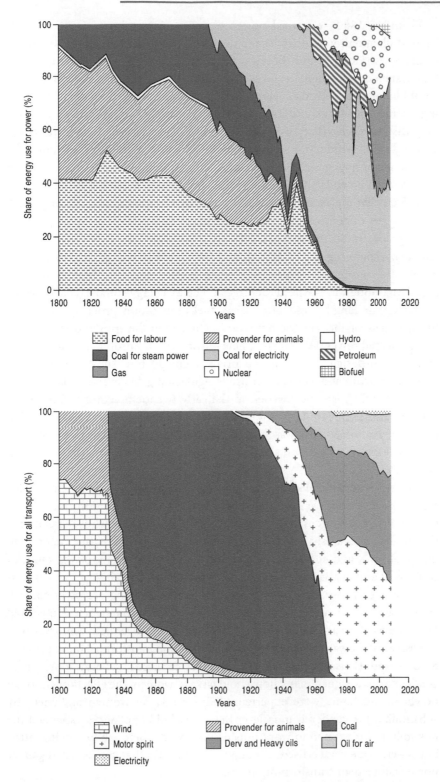

Figure 9.5 Transitions in power and transport services, UK 1800–2020 (based on Fouquet 2010)

From the 1880s, electrical power played an increasingly significant role beyond illumination. Electricity rapidly developed new markets across a range of energy services, in part because the difficulty of storing electricity meant that system development required nurturing sufficient demand. Enabled by a combination of technical developments, including motors that ran on alternating current and transformers that made it possible to transmit power over long distances, electricity ushered in a wave of transformations in transport and industrial production. This 'second industrial revolution' included the production of new materials (e.g. aluminium) and industrial techniques (such as electroplating and galvanisation) whose production relied on abundant, low-cost electricity; replacement of steam power by electric motors, offering much finer degrees of control and precision; and communication technologies such as telephony, radio and radar. Electricity transformed the experience of using energy by 'liberating' the energy content from material energy carriers like coal or oil, and serving it up precisely when needed and in a highly controllable form. Gone was the need for large and bulky energy storage (coal bunkers, oil tanks), or lengthy and often dirty fuel preparation processes. Many of the materials of the 'electric age' were lighter than the iron and steel of the first industrial revolution, creating new possibilities for product, industrial and architectural design. Economic historians highlight how substituting electric motors for steam and waterpower drove significant gains in industrial productivity in the UK, US and Sweden in the early twentieth century (Enflo et al. 2009; David 1990). Similarly, the mid-twentieth century dominance of the US in global manufacturing has been attributed to its "unsurpassed access to affordable energy in general and to inexpensive electricity in particular" (Smil 2010a: 67).

Electric motors enabled a reduction in the size and weight of machinery for a given power output. Within factories, this led to a multiplication and redistribution of power sources away from a single central engine whose operational rhythm had previously governed the pace of work for the entire factory. On farms, at work and at home, it drove the motorisation of a growing range of equipment and appliances. For the end consumer, electrical power was experienced as a 'dematerialisation' of energy use and a growing degree of control, epitomised by the supreme convenience of the flick of a switch. Household applications of electricity rapidly expanded in the mid-twentieth century: irons, stoves, vacuums, washing machines and refrigeration transformed the experience of, and aspirations for, domestic life (Kander et al. 2013). Because prevailing gender norms assigned many household tasks to women, the effects of electrical power in the home were experienced differently by women and men. By substituting for gendered muscle-power, household appliances expanded the possibilities for women to work outside the home. Their 'labour-saving' attributes were widely advertised, although often in ways that consolidated gender norms rather than transforming them.

Transport

How people and goods move around has changed markedly over the past two centuries. Costs for land and sea transport have fallen and distances travelled per person have grown, so that long-distance travel (of both people and things) is now the norm for a large portion of the world's population. Entirely new transportation experiences have emerged, such as air travel and the possibility to move at speeds greater than a galloping horse: as the novelist and philosopher Aldous Huxley observed, speed is perhaps the only uniquely modern experience. Behind these profound changes in spatial experience are a complex set of transitions that include the mechanisation of transport, development and funding of dedicated infrastructure (canals, railways, roads), and shifts towards progressively higher quality fuels (Figure 9.5). By decreasing the friction of movement and making available much greater mobile power, these shifts have increased speeds and/or dramatically driven down the cost of moving across space.

Several waves of innovation and investment had substantially improved animal-powered transportation by the early nineteenth century. These included the development of canal networks, major programs of road improvement (such as the construction of turnpikes and river crossings), innovations in coach and wagon building, and techniques of animal breeding to optimise animals for either power (like the shire horse) or speed. The nineteenth century, however, saw a decisive shift in transportation towards the application of heat for mobile power. This first took the form of the steam engine using coal as a heat source, although wood and oil also fuelled steam engines in the nineteenth century, particularly where coal was scarce: wood was the primary fuel for steam boats and railroads in the United States until the 1850s, for example, and its use in the western states continued until much later (Melosi 1982). The application of steam power to locomotion required the perfection of high-pressure steam, which concentrated available power and reduced engines to a manageable size. This process took nearly a century following Newcomen's introduction of steam for stationary power (1712): the first steam-powered road vehicle debuted in 1801, Richard Trevithick's self-powered engine was the first application of steam to rail in 1804, and steam was in commercial use in the marine sector by 1818 (Kander et al. 2013). On water, horse-drawn barges and sailing ships gradually gave way to steam during the nineteenth century, although sailing ships hung on into the twentieth century as competition from coal spurred innovation in sail and ship design. On land, animal power (horses and oxen) ceded dominance to steam for long-distance journeys and large loads.

The transition to steam-power (and from wood to coal) was an evolutionary process, but its effects were revolutionary (Melosi 1982). By increasing speeds and reducing the cost of transportation, railways fundamentally transformed the experience of space and time. The requirements of timetabling a spatially extensive network led to the introduction of standardised (national) time zones, replacing a multitude of local times. Railways were especially influential as a transportation infrastructure because of their much greater geographical reach than canals and early road networks. By dramatically reducing the costs of transporting bulky low-value goods (such as coal), railways consolidated the transition to steam power in places that were distant from local coal sources, coasts or river transport so that "the use of coal as fuel could become a continent-wide phenomenon" (Kander et al. 2013: 192). Rail travel propelled a form of 'time-space compression' (**Chapter 2**) that brought cities on the rail network closer together in terms of travel time and functional integration but, in doing so, also increased the relative distance between these cities and other places not connected to the railway. Overall, energy transitions in transport have created new patterns of uneven development rather than flattening spatial difference.

A growing proportion of land and marine transport moved away from steam power towards the oil-fuelled internal combustion engine in the early twentieth century. A functional internal combustion engine had emerged as early as the 1870s but its application to road transportation was restricted to the very wealthy (Fouquet 2010). Cost reduction in the early twentieth century via mass production, combined with the mass-marketing of automobiles, led to a re-appraisal of the 'old' infrastructure of roads, and their re-purposing (widening, straightening, resurfacing, strengthening) for motorised vehicles. The power-to-weight advantages of the internal combustion engine over steam power, higher speeds, and relative ease of fuelling with liquid oil vs solid coal led to its rapid uptake in the marine sector. Early marine diesels were in service in Europe in the first decade of the twentieth century, the British Navy switched from coal to oil by the end of the First World War (with the world's other navies quickly following suit) and, by the middle of the century, diesel power was dominant in shipping. Today, large marine diesel engines power the container ships and tankers of global trade: fuelled by heavy 'bunker' oil (emissions from which are not directly included in the Paris Climate Change Agreement), these workhorses of globalisation can drive a 9-metre propeller at around 100 rpm, moving a 18,000 box container ship at over 40 kph (Smil 2010b). Although the average speed of marine transport has changed relatively little, the primary effect of the transition to oil (as with the transition from sail to steam) has been sharp reductions in the cost of transportation, and a corresponding increase in the scale of movement. A similar transition from steam power to heavy diesel engines also took place on the rails. Experiments with diesel-electric rail locomotives in the 1920s overcame basic problems of power transmission and, by the 1950s, the greater flexibility and performance of diesel engines had encouraged their replacement for steam in most countries, except where coal was plentiful and oil limited in supply (such as South Africa, India and China). In general, coal had exited the transportation sector as a fuel by the 1970s, to be replaced by petroleum products (in shipping, road transport, rail, air) and electricity (primarily high-speed rail, and suburban commuter lines).

Railways provided the first novel taste of speed, but air travel took this to the next level. Successively faster propeller driven aircraft emerged in the years running up to the Second World War. Air travel was expensive, however, and limited to a niche markets as "providing no cargo services and generating few mail contracts, there was nothing to cross-subsidise passenger transport" (Fouquet 2008: 183). The development of the jet engine transformed the possibilities for speed, and the first jet engines entered commercial service in 1952. Subsequent improvements to engine efficiency have enabled larger loads to be carried, opening up the possibility of cheaper air travel for both long and short-haul routes, and the opportunity for a large air-cargo market to emerge. Air travel experienced very rapid growth in the second half of twentieth century and there has been an even sharper rise in air cargo transport. The gas turbine engines developed to power jet airliners have subsequently diffused beyond the aviation sector and, like steam engines and electric motors before them, can now be considered a general purpose technology. Gas turbines have some fast maritime transport applications (including military and vehicle ferry uses) but their most extensive use is for the generation of stationary power. Powered by natural gas or diesel fuel, gas turbines have become a staple feature of electrical power generation.

With its beginnings in illumination, electricity had emerged at the end of the nineteenth century as an important means of mass urban transport alongside the dominance of steam power. The extension of electric railways and tramlines powered waves of early twentieth century real estate development on the urban fringe: the world's first underground electric railway opened in London in 1890 and, within twenty-five years, similar infrastructures were in place in North America (Boston, New York City, Philadelphia), Europe (Athens, Berlin, Budapest, Glasgow, Hamburg, Liverpool, Paris) and South America (Buenos Aires). Electric cars were put into public service as taxis in major cities in North America and Europe at the end of the nineteenth century: in 1900, 38% of vehicles in the US were powered by electricity, 40% on steam and 22% on gasoline and the electric vehicle fleet continued to grow in the early years of the twentieth century. By 1930, however, electrically propelled cars had been largely displaced by internal combustion engines in the US and Europe. In some places electricity also exited the urban mass-transit market: between 1936 and 1950, electrically driven tramcar lines in over 45 major American cities were purchased by an alliance of automobile and oil companies and shut down, leaving oil-powered transportation in their place (Bridge and Le Billon 2017).

CONCLUSION

Energy transitions are major structural shifts within an energy system. In this chapter we have identified a number of historical energy transitions at household, urban, national and/or global scales. Research on past energy transitions has focused on evolutionary shifts in the role of different primary fuels and conversion technologies in the energy mix, such as the transition from wood and water power to coal in the nineteenth century, or from coal to oil in the twentieth. There is a lot to be learned from these past transitions, and the way they have been associated with broad and often far-reaching social change. Energy transitions have propelled the development of novel energy landscapes such as deep coalmines, town gas works, petrol stations and household kitchens saturated with electrical appliances, and the abandonment of others (**Chapter 1**). By creating new ways to deliver heat, light, transport and power and transforming the costs of these energy services, energy transitions have influenced the rise and fall of different economic sectors and contributed to new geographies of uneven development (**Chapter 2**). Shifts in energy carriers, technologies and infrastructures have also forged new connections between different places and communities, as they become bound together within expanding networks of energy infrastructure (see **Chapter 3**).

Energy transitions have also transformed sensory experiences in significant ways, including such everyday experiences as indoor air temperatures, urban air quality, soundscapes, speed and acceleration. Enormous scale shifts in available power and mobility have transformed the social meaning and significance of space and time. Paradoxically, the same time energy itself has become increasingly invisible within spaces of consumption. Some consequences of past transitions are less obvious but no less significant. Amongst the most far reaching are the political identities fostered around different energy systems (**Chapter 4**), particularly those associated with spaces of work, consumption and pollution. The vast number of people needed to extract coal and move it by rail and water, for example, and the dense geographies of coal mining work and the communities it supported, created conditions for collective political organising and socio-political struggle. These conditions transformed the energy bottleneck held by miners and other coal workers into political power, expressed, for example, in the role of coal workers in the labour movement, or the capacity of striking miners to bring down national governments (Mitchell 2011). The same conditions do not hold for oil (which also requires far fewer people to extract and mobilise an equivalent amount of energy), and so coal's declining dominance in the overall energy mix within many countries has eroded a space of collective political action that has been important in shaping the welfare state in the twentieth century.

The energy transitions we have identified in this chapter have also shaped the experience of mass consumption in significant ways. Long-run declines in the price of energy services have enabled people on relatively modest wages to access a range of experiences (abundant light, home ownership, leisure time, family vacationing), accumulate goods (automobiles, household consumer durables), and maintain living conditions (heating and cooling, practices of food provisioning) that could once only be procured at great cost. At the same time, documentation and testimony of the negative effects of energy systems – from poor working conditions, urban air quality and radioactive waste disposal – have driven powerful political movements for justice and reform. In summary, past energy transitions have been about much more than the substitution of fuels according to the grand fuel sequence, or the innovation and diffusion of energy conversion technologies. As we outline in **Chapter 10**, a low-carbon energy transition is likely to be as significant – and its social, technological and geographical implications as hard to imagine – as the shift from wood to coal, or the electrification of urban and rural areas in the late nineteenth century.

QUESTIONS FOR DISCUSSION

- Consider a city with which you are familiar: what evidence is there in the built environment for historic shifts in how energy is used? Are there other significant shifts that have occurred for which there is little/no evidence?

- How are household-scale changes in energy provisioning during the twentieth century related to changes in energy use at other spatial scales?

- How has the geography of pollution (associated with energy consumption) changed as a result of changes in urban energy use?

- How might testimonies to the transformative experience of historic energy transitions at the houschold scale (such as the advent of electricity) inform the design of future energy transitions?

ACTIVITIES/POTENTIAL RESEARCH PROJECTS

- Examine how one energy service (e.g. heat, light or power) has been delivered over time in a country/city of your choice. What have been the political implications of these changes in energy use?

- Consider commonalities and differences among three historic energy transitions: what can these historical energy transitions tell us about the distinctiveness of a low-carbon transition?

RECOMMENDED READINGS

■ Fouquet, R. 2016. Historical energy transitions: speed, prices and system transformation. *Energy Research & Social Science* 22: 7–12.

This article presents evidence on the speed and driving forces of historic energy transitions. It argues that the price of energy services to the consumer plays a crucial role in creating incentives that stimulate energy transitions (as consumers like a bargain), but that price shocks can also stimulate this process and catalyse system change. In addition, new technologies that offer greater value to the consumer (cleaner, easier) – even if the initial price is higher – can drive transition. The author shows how energy transitions can be held up by the reaction of incumbent and declining industries. The article concludes that the current fossil fuel energy system needs to be a springboard from which an alternative can emerge, but that energy transitions are not inevitable: they depend on actors and forces of path creation and often involve political will, to create demand for alternatives and address the power of incumbent sectors.

■ Kander, A., P. Malanima and P. Warde. 2013. *Power to the people: energy in Europe over the last five centuries.* Princeton, NJ, and Oxford: Princeton University Press.

A trove of information on the relationship between energy consumption and economic development in Europe over five centuries written by three economic and environmental historians. The book is more than a descriptive compendium, however, and systematically examines the drivers and consequences of the energy transitions associated with the 'three industrial revolutions' (coal and steam, electricity and oil, and information technology).

■ Needham, A. 2014. *Powerlines: Phoenix and the making of the modern south-west.* Princeton, NJ: Princeton University Press.

This book joins together, via the medium of electricity, the process of suburban growth in the American Sunbelt and the post-war transformation of the Colorado Plateau into a major centre of coal mining and power generation. Adopting a regional (rather than urban) approach to the process of transition in energy systems enables the author to show how transitions in energy supply and demand took place through the integration of different geographies. New energy experiences (refrigeration, air conditioning) and norms of consumption in places like Phoenix, for example, depended on a rapid transition to coal-fired power generation starting in the 1950s. It involved the opening of large mines and power plants on the Colorado Plateau, and development of high voltage power lines to transfer electricity to distant urban centres.

■ Sovacool, B. 2016. How long will it take? Conceptualizing the temporal dynamics of energy transitions. *Energy Research & Social Science* 13: 202–215.

This article also examines the important question of time in energy transition. It reviews the conventional wisdom that suggests that transitions are long drawn-out affairs that can take somewhere between 50–70 years (for a new fuel source or prime mover) and even longer for a transition involving all economic sectors and energy services. The author then presents ten counter examples which indicate much more rapid energy transitions are possible. These have involved changes in end-use energy devices (e.g. lighting, air-conditioning, cookstoves, ethanol powered vehicles) and shifts in energy supply (e.g. introduction of combined heat and power in Denmark, nuclear power in France, phase out of coal in Ontario, switch to natural gas in the Netherlands) and have taken, on average, around a decade or so. The article concludes by summarising the policy implications of this counter-evidence on the pace of transition.

REFERENCES

Allen, R. 2012. Backward into the future: the shift to coal and implications for the next energy transition. *Energy Policy* 50: 17–23.

Allen, R. 2013. The shift to coal. In RCC Perspectives, No. 2, *Energy Transitions in History: Global Cases of Continuity and Change*. R. Unger (ed.), Munich: Rachel Carson Center, pp. 11–16.

Baker, L., P. Newell and J. Phillips. 2014. The political economy of energy transitions: the case of South Africa. *New Political Economy* 19(6): 791–818.

Basalla, G. 1982. Some persistent energy myths. In *Energy and Transport: historical perspectives on policy issues*. G. Daniels and M. Rose (eds), Beverly Hills: Sage, pp. 27–38.

Bernton, H., W. Kovarik and S. Sklar. 2010. *The forbidden fuel: a history of power alcohol*. Lincoln/London: University of Nebraska Press.

Bordoff, J. 2017. America's energy policy – from independence to interdependence. Available online at www.cirsd.org/en/horizons/horizons-autumn-2016—issue-no-8/americas-energy-policy-from-independence-to-interdependence

Bridge, G., and P. Le Billon. 2017. *Oil.* Cambridge: Polity Press (2nd edition).

David, P. 1990. The dynamo and the computer: an historical perspective on the modern productivity paradox. *The American Economic Review* 80(2): 355–361.

Enflo, K., A. Kander and L. Schön. 2009. Electrification and energy productivity. *Ecological Economics* 68 (11): 2808–2817.

Fouquet, R., 2008. *Heat power and light: revolutions in energy services*. Cheltenham and Northampton, MA: Edward Elgar Publications.

Fouquet, R. 2010. The slow search for solutions: Lessons from historical energy transitions by sector and service. *Energy Policy* 38(11): 6586–6596.

Fouquet, R. and P.J. Pearson. 2006. Seven centuries of energy services: The price and use of light in the United Kingdom (1300–2000). *The Energy Journal* 27(1): 139–177.

Fouquet, R. and P.J.G. Pearson. 2012. Past and prospective energy transitions: insights from history. *Energy Policy* 50: 1–7.

Gailing, L. and T. Moss, (eds) 2016. *Conceptualizing Germany's energy transition: institutions, materiality, power, space*. London: Springer.

Geels, F. 2005. *Technological transitions and system innovations: a co-evolutionary and socio-technical analysis*. Cheltenham: Edward Elgar.

Hecht, G., 1998. *The radiance of France: nuclear power and national identity after World War II*. Cambridge, MA: Massachusetts Institute of Technology.

Hirsh, R. and C. Jones. 2014. History's contributions to energy research and policy. *Energy Research & Social Science* 1: 106–111.

Holl, J. 1982. The Nixon sdministration and the 1973 energy crisis: a new departure in federal energy policy. In *Energy and transport: historical perspectives on policy issues*. G. Daniels and M. Rose (eds), Beverly Hills: Sage Publications, pp. 149–156.

Hughes, T. 1993. The evolution of large technical systems. In *Social construction of technological systems: new directions in the sociology and history of technology*. W. Bijker, T. Hughes and T. Pinch (eds), Cambridge, MA: MIT Press, pp. 51–82.

Jasanoff, S. and S.H. Kim. 2009. Containing the atom: sociotechnical imaginaries and nuclear power in the United States and South Korea. *Minerva* 47(2): 119.

Johnson, V., F. Sherry-Brennan and P. Pearson. 2016. Alternative liquid fuels in the UK in the inter-war period (1918–1938): insights from a failed energy transition. *Environmental Innovation and Societal Transitions* 20: 33–47.

Kander, A., P. Malanima and P. Warde. 2013. *Power to the people: energy in Europe over the last five centuries.* Princeton, NJ, and Oxford: Princeton University Press.

Kennedy, L. 2013. 'The people's fuel': turf in Ireland in the nineteenth and twentieth centuries. In *Energy transitions in history: global cases of continuity and change.* R. Unger (ed.), Munich: Rachel Carson Center, pp. 25–30.

Krausmann, F. 2013. The social metabolism of European industrialisation: changes in the relation of energy and land use from the eighteenth to the twentieth century. In *Energy transitions in history: global cases of continuity and change.* Unger, R. (ed.), Munich: Rachel Carson Center, pp. 31–36.

Malm, A., 2016. *Fossil capital: The rise of steam power and the roots of global warming.* London: Verso Books.

Markusson, N., M.D. Gjefsen, J.C. Stephens and D. Tyfield. 2017. The political economy of technical fixes: a case from the climate domain. *Energy Research & Social Science* 23: 1–10.

Melosi, M. 1982. Energy transitions in the nineteenth century economy. In *Energy and transport: historical perspectives on policy issues.* G. Daniels and M. Rose (eds), Beverly Hills: Sage, pp. 55–70.

Mitchell, T. 2011. *Carbon democracy: political power in the age of oil.* London and New York: Verso Books.

Rutter, P and J. Keirstead. 2012. A brief history and the possible future of urban energy systems. *Energy Policy* 50: 72–80.

Shove, E., G. Walker and S. Brown. 2014. Material culture, room temperature and the social organisation of thermal energy. *Journal of Material Culture* 19(2): 113–124.

Simmons, I. 2008. *Global environmental history: 10,000 BC to AD 2000.* Edinburgh University Press.

Smil, V. 2005. *Energy at the crossroads: global perspectives and uncertainties.* Edinburgh: Cambridge, MA: MIT Press.

Smil V. 2010a. *Energy transitions: history, requirements, prospects.* Santa Barbara, CA: Praeger.

Smil, V., 2010b. *Prime movers of globalization: the history and impact of diesel engines and gas turbines.* Cambridge, MA: MIT Press.

Turnheim, B. and F. Geels. 2012. Regime destabilisation as the flipside of energy transitions: lessons from the history of the British coal industry (1913–1997). *Energy Policy* 50: 35–49.

Wallsten, B. 2015. *The Urk World: hibernating infrastructures and the quest for urban mining.* Dissertation, Linköping University Electronic Press.

Wrigley, E.A. 1990. *Continuity, chance and change: the character of the industrial revolution in England.* Cambridge: Cambridge University Press.

Future transitions

10

LEARNING OUTCOMES

- Understand the nature of energy transitions as a political and economic project and the implications for analysing who benefits and who loses from transition processes.
- Critically explore different visions for energy transitions, and how and why these vary with social and geographical contexts.
- Explore how transition processes take place and evaluate different approaches for analysing the transition process and their strengths and limitations.
- Critically examine the role of innovation and experimentation in energy transitions, and the consequences for how we understand the politics of energy transition.
- Reflect on how energy geographies evolve over time and the dynamics of development, innovation and contestation that shape energy futures.

As we have illustrated throughout this book, energy has been fundamental to the development of societies, economies and their politics, and to the evolution of the **socio-technical regimes** and everyday practices which they both sustain and challenge. If energy is essential to understanding the historical and geographical emergence of contemporary societies, it is equally vital to the ways in which their futures are imagined and pursued. Through each of the chapters, we have examined how the four imperatives of providing secure, affordable and sustainable energy societies in ways that are socially just (what we have termed the *Energy–Society Prism*, see **Introduction**) is manifest in different contexts as different groups seek to promote particular energy technologies, policies and interests. Yet, we have also seen how challenging it can be to change energy systems as innovations fail and incumbent interests seek to maintain the status quo.

socio-technical regime
the stable arrangements of social and technical elements held in place by infrastructures, formal rules and informal conventions, which serve to reproduce energy systems.

In **Chapter 9**, we showed how the idea of transitions helps us understand how, why and when energy systems undergo change. We explored the ways in which energy systems have evolved over time, changed both by the actions of individual entrepreneurs and by powerful political and economic interests. Such historical transitions in the ways in which we provide and use energy have often been incremental, taking place over the *longue durée* seemingly with little orchestration or intent. At the same time, it is possible to identify more purposive attempts to design or reconfigure energy systems, such as development of nuclear power or the turn to smokeless fuels in many industrial cities in the post war period. As energy once again rises to the top of the political agenda, policymakers, businesses and communities are engaged in a new wave of interventions intended to forge new energy futures. However, as with past energy transitions, such efforts are far from uniform or coherent. Not only do visions of future energy systems vary significantly, but the realisation of specific projects and programmes is shaped by the social and geographical contexts through which they emerge.

In this final chapter, we explore the visions and interventions that are being developed in order to realise energy futures and assess their implications. The first half considers the different visions for energy systems that currently dominate public and political discourse – those focused on a **low-carbon economy**, those concerned with ensuring access to energy for development, those promoting the continued development of fossil fuel energy resources, and alternative approaches that advocate the reduction in the use of energy. These visions are giving rise to important political projects, from the large-scale reconfiguration of energy systems, to the opening up of new fossil fuel reserves, from the emergence of new technologies to the development of alternative ways of living. While each of these forms of energy transition is visible, their extent, impact and implications vary from sector to sector. The second half of the chapter turns to examine how the transformation of energy systems is achieved in practice. Rather than being a matter of policy change or the right economic incentives, research has pointed to the ways in which managing **niche innovations** is critical to achieving energy transitions. We explore how forms of innovation and **experimentation** are taking place at the urban level in response to the vision of low-carbon energy futures. We conclude the chapter by reflecting on the main themes of the book and considering their implications for our energy futures.

low-carbon economy
a term used to denote forms of economic activity that are able to generate employment and profit through the pursuit of low-carbon transitions, such as the development of a renewable energy manu-facturing sector or forms of the service economy required to support transitions.

niche innovations
niches are spaces protected from incumbents in which novel socio-technical configurations can become established; the development of niches, and the innovations they enable, has been identified by researchers as a key pathway to transition (compare to destabilisation, **Chapter 9**).

experimentation
a process of learning by doing in which novel technical entities or social arrangements are tried out, tested, or piloted.

WHICH ENERGY TRANSITION?

Energy's ubiquity, multiplicity and necessity lend it a political malleability unmatched in other areas of resource politics. Strategic political and economic interests together with the priorities given to the challenges of energy security, affordability and sustainability in different socio-economic contexts mean that there are multiple and competing visions of which energy transition is most desirable and how it should be achieved. This malleability means that energy futures can be envisaged alternatively as low carbon or fossil fuel abundant; as smart and efficient or wasteful; as unlimited or finite. Yet, amid the clamour of claim and counter-claim, distinct **narratives** are emerging that seek to shape the trajectories of energy systems in particular directions. Of particular importance has been the growing recognition of the importance of achieving a *low-carbon transition* in energy systems in order to reduce their contribution to atmospheric concentrations of greenhouse gases. Accompanying this has been a growing movement seeking to promote the importance of *energy for development*. Yet, at the same time, and with gathering momentum around the discovery of new resources, *fossil fuel futures* continue to be advocated as essential to achieving secure and affordable energy systems. In contrast, other narratives suggest the need for *reduced consumption* in order to ensure the future of increasingly scarce resources and promote sustainability. Such narratives are vital for

narratives
discourses, visions and rationales that link together various elements in order to tell a coherent story about the way in which we should understand the world. They are powerful means of ordering both present and futures societies.

Figure 10.1 Solar power in the city: multiple visions of renewable energy futures in Cape Town and Stellenbosch, South Africa (photo credit: Harriet Bulkeley)

understanding both the present and future of politics of energy, for "climate and energy policy discourses do not simply describe human relationships with material artifacts like coal-fired power plants, but actively structure these relationships" (Rafey and Sovacol 2011: 1142). As Figure 10.1 shows, these narratives extend beyond the realm of energy and connect to wider societal concerns. In this section, we consider each of these visions in turn, exploring how, by whom and towards what ends they are being articulated. From this review, we can see that these visions are articulated in diverse ways across energy sectors, with narratives concerning the *low-carbon* transition and the importance of *energy for development* being most prevalent in relation to the electricity sector, while the continued use of *fossil fuels* continues to be dominant in relation to transport.

Low-carbon transition

The notion of a low-carbon transition, or the need to develop the technologies, systems, and policies through which to develop and grow an economy that would be based on low-carbon energy resources, has grown in popularity through the first part of the twenty-first century. Such is the prevalence of the term low carbon that the origins of the notion of a low-carbon transition are hard to pin down. One important reference point was the report by the Performance and Innovation Unit of the UK's Cabinet Office in 2002 which suggested that "the UK should be setting about creating a range of future options by which low carbon futures could be delivered" (PIU 2002: 9). The 2003 Energy White Paper, *Our Energy Future: creating a low-carbon economy* (DTI 2003) drew heavily upon this report and the earlier work of the Royal Commission on Environmental Pollution to argue that meeting new energy challenges provided

> the opportunity to shift the UK decisively towards becoming a low-carbon economy where higher resource productivity – producing more with fewer natural resources and less pollution – will contribute to higher living standards and a better quality of life.
>
> (DTI 2003: 6)

Following the 2008 Climate Change Act that created binding carbon budget targets for the UK to meet, the 2009 Energy White Paper continued to champion the need for the UK to become a "low-carbon" country. Guided by ambitious targets to reduce greenhouse gas emissions by 60% by 2050 and 80% by 2080, the UK's energy policy contains multiple measures and interventions, from supporting the development of a second generation of nuclear power stations to the Green Deal scheme to encourage energy efficiency improvements in the built environment. Here, the notion of a low-carbon transition has developed in a relatively piecemeal fashion, diffusing controversy through embracing multiple interpretations of what it might mean to be low carbon (Lovell et al. 2009).

The notion of a low-carbon transition has evolved in a starkly different manner in Germany. In 2010 an ambitious strategy for the long-term decarbonisation of the German economy was released, following Germany's long-standing commitment to addressing climate change and the growing momentum across the OECD surrounding the idea of the low-carbon economy. Yet in 2011, in the wake of the Fukushima nuclear disaster in Japan, "German politicians – led by chancellor and physicist Angela Merkel – abruptly shut down eight of Germany's 17 nuclear plants and, several months later, decided to completely phase out nuclear energy by 2022" (Stegen and Seel 2013: 1481). The **Energiewende** as it has been termed, requires Germany to meet new targets for reducing energy use and developing the role of renewable energy in the transition to a low-carbon economy (see **Example 1**). The emergence of the notion of a low-arbon transition has had far-reaching effects beyond Europe. The EU has been a strong advocate for *Low Emissions Development Strategies* under the auspices of the UNFCCC such that "the concept has been included in the negotiating texts ... since the run up to COP15 in Copenhagen in 2009 and is part of both the Copenhagen Accord" (UNFCCC, 2009) and the Cancun Agreements (UNFCCC, 2011), which recognize that a LEDS is indispensable to sustainable development and that incentives are required to support the development of such strategies in developing countries" (UN 2018).

Beyond the actions of national governments, the discourse of low-carbon transitions has also become ubiquitous amongst sub-national and city governments, private sector organisations and communities (Bulkeley and Newell 2015; Bulkeley et al. 2010). HSBC, for example, developed an internal climate change strategy that saw them become the first carbon neutral bank in 2005 and have since developed what they term their 'climate business', focused on developing and delivering the financial investment required for the low-carbon economy (Reference: http://www.hsbc.com/our-approach/sustainability/finance).

Energy for development

The narrative of the low-carbon transition has also been taken up by those advocating the need to improve access to secure and affordable energy services. Providing access to "affordable, reliable, sustainable and modern energy for all" is one of the Sustainable Development Goals adopted by the United Nations in 2015. The provision of energy to support economic development has historically been an important driver of how national governments have sought to design and develop energy systems. This historical concern has mainly been focused on ensuring that there is sufficient energy (particularly electrical power) to support the development of critical economic sectors – for example, the mining industry in South Africa. International momentum for addressing energy and development has been driven by wider concerns about the lack of access to energy services experienced by many of the world's population and a more encompassing idea of economic development. From this growing international

---

EXAMPLE 1: THE ENERGIEWENDE

After two decades of ambitious climate change and renewable energy policies, in 2010 Germany's federal government presented its most ambitious energy strategy yet. Its goals were to reduce CO_2 emissions by 40% by 2020, 55% by 2030 and 95% by 2050 compared with 1990 levels. Rapid reductions in GHG emissions had been achieved early on, in part, because of the historical circumstance of using 1990 as a baseline year (the figure for baseline emissions was inflated because it included many inefficient coal-fired power stations in the former Eastern Germany that were quickly replaced). Nonetheless, these new targets represented a significant challenge. At the time, the German government argued that nuclear power would be required as a 'bridging fuel' in order to meet these ambitions. However, the strong reaction to the Fukushima nuclear accident in Japan in 2011 saw a reversal of this position (see **Chapter 9**). Nuclear energy represented just under a quarter of German electricity supply in 2011, but the decision to immediately close eight nuclear plants with the nine remaining to be phased out between 2015 and 2050 met with cross-party support. In place of nuclear energy, renewables will provide 60% of Germany's energy.

The enormity of the challenge has not been lost on policymakers or the energy industry, and critics have labelled the Energiewende (literally 'energy turn' in German) as too expensive, too fast, and too ambitious. Key challenges include reconfiguring the systems of energy distribution to enable multiple sources of electricity from different kinds of renewables to be used, energy storage, the incumbent roles of the big four energy utilities (RWE, E.ON, Vattenfall and EnBW), and the role of the state (Moss et al. 2015: 1549). At the same time, while the Energiewende is a national policy, its realisation is heavily dependent on the actions of communities, municipalities and regions. As Tim Moss and his colleagues (2014: 2) point out, some "prominent examples include energy cooperatives (around windfarms, solar parks, etc.), so-called bio-energy villages and regions (aspiring to 100% reliance on renewables) and initiatives to re-municipalise energy utilities and/or electricity networks". Early research suggests that the fortunes of such initiatives are mixed, with the proposed transition to renewable energy far from assured. Despite the strategic nature of the national policy intervention that gave rise to the Energiewende, in practice its realisation depends on the reconfiguration and alignment of multiple socio-technical configurations under conditions of political contestation. This case suggests that transitions to future energy systems, even where powerful visions and interventions are present, are likely to evolve in a more fragmented and contested fashion and lead to multiple unexpected outcomes.

---

development perspective, a lack of access to energy services is seen as a critical factor shaping educational attainment and public health, as well as shaping the potential for economic development amongst communities. The emergence of the idea of 'sustainable and modern energy for all' has both been driven by the recognition of the environmental issues associated with traditional forms of energy provision for development, which have tended to rely on fossil fuels. At the same time, the interest in the potential for renewable energy has been

CASE STUDY: LIGHTING A BILLION LIVES (LaBL)

Contributed by Ankit Kumar, Eindhoven University of Technology, the Netherlands

LaBL aims to provide electric lights through solar lanterns and solar micro-grids to a billion people across the world. With its partners, typically local NGOs, LaBL identifies villages that need its intervention and a village entrepreneur who is responsible for day-to-day operations in the village. A solar lantern charging station, which typically charges 50–60 solar lanterns, is set up in the entrepreneur's home. In India, lanterns are rented to villagers for around 3–5 Indian rupees per day, approximately £0.04 or US$0.06. Villagers need to deposit the lanterns every morning for charging and pick them up for use every evening. In this way, LaBL not only provides lighting services but also creates employment in the villages. In addition, it trains entrepreneurs and local NGO staff in operation and maintenance of the solar lanterns and charging stations. LaBL argues that it contributes to gender empowerment by giving priority to women and youth when it comes to selecting village entrepreneurs.

Solar projects are funded through sponsorship from industries, corporate houses, celebrities and individuals. Sponsors make commitments to 'light a village' or multiple villages. LaBL has also received financial support from the Ministry of New and Renewable Energies (MNRE), Government of India. Every year, a national news channel in India (NDTV) runs telethons with a leading think-tank, The Energy and Resource Institute (TERI), to drum up support and funding for LaBL. These telethons last for 24 hours and feature several television and film celebrities dancing and singing for a 'good cause'. Celebrities often advertise their upcoming films and corporate houses advertise their products on the telethons, which also feature villages, homes and lives that have been 'lit' by LaBL. Sponsors typically visit these villages and interact with the communities, focusing on the ways in which lives have been changed by the presence of light. In return for sponsorship, villagers are expected to open their homes and lives to scrutiny by television cameras and sponsors.

LaBL has a presence in 2580 villages covering nearly 450,000 households, and has brought solar lanterns and micro-grids to over 2 million lives (TERI n.d.). It argues that solar lanterns mobilise several activities that were either difficult or impossible due to the problems associated with kerosene lamps. There has been an improvement in the health of children who were previously exposed to health hazards from inhaling fumes from kerosene lamps. Furthermore, the better quality light from solar lanterns also makes it easier for children to study. LaBL provides livelihood opportunities by creating jobs and the solar lanterns also help local businessmen and shopkeepers stay open for longer as they no longer have to worry about fuel costs. However, the benefits claimed by LaBL do not reach everyone. Since there are only 50–60 lanterns per village, not everyone in a typical village of 100–200 households gets direct access to the solar lanterns. Those who get access are mostly from within the entrepreneur's social network and issues like caste and class often play a role. The solar lantern can be used to light up only one space in the household at a time, unlike kerosene that can be distributed in several lamps and used to light multiple spaces at the same time. Thus, rather than replacing kerosene lamps, in most households solar lanterns share space with kerosene lamps. While this may suggest increased access to energy services, it can come at a cost households are unable to afford.

Figure 10.2
LaBL's solar programme, India (clockwise from top
left): solar lantern charging station with female
entrepreneur; solar lantern being used during tuition
in a village; a solar lantern in a kitchen where a child
is also studying (photo credit: Ankit Kumar)

shaped by the interest in providing energy access 'on the ground' and a wider notion of development that encompasses education, health, well-being and community empowerment.

Reflecting this switch in how energy and development are coupled, low-carbon and renewable energy systems have come to be central to many development projects. Partly funded by new forms of climate finance but also supported through traditional aid donors and development agencies, such projects take multiple forms, from the provision of cookstoves that either burn traditional wood-fuels more efficiently or use alternative resources, to the retrofitting of low income housing, to the provision of renewable energy at multiple scales. Lighting a Billion Lives (LaBL) is one such example (see **Case study**). Started in 2008 as Lighting a Million Lives when an Indian NGO, The Energy and Resources Institute (TERI) made a pledge "to bring light to one million people in India by replacing kerosene and paraffin lanterns with solar power lighting devices" at the Clinton Global Initiative meeting (Clinton Foundation, 2008). Figure 10.2 illustrates some of the ways in which this scheme is being implemented on the ground.

Fossil fuel futures?

Alongside the growing momentum of visions of low-carbon energy futures, fossil fuels remain deeply entrenched in the ways in which energy futures are imagined and designed. First, governments and industries globally continue to invest in the development of traditional fossil fuel reserves, from coal to oil and natural gas. In this narrative, the imperatives of national development and economic growth are regarded as outweighing any concerns for sustainability – or at least to displace such concerns geographically, to other nation-states who are deemed to have more responsibility or capacity to do so. One example of where energy futures are framed in terms of the continued development of fossil fuel resources is South Africa. Here, "cheap, plentiful energy is framed as crucial to fuel continued economic growth, help South Africa recover from the global crisis, and bolster industrial competitiveness", providing the basis for investment in the world's largest dry-cooled coal-fired power station in the world, the 4,800 MW capacity Medupi situated in the Limpopo province in the North-East of the country (Rafey and Sovacol 2011: 1141). Funded by a loan from the World Bank of $3.75 billion, the plant represents the single largest investment by the South African power utility Eskom and is designed to provide 10% of the country's electricity, while also generating "more emissions – 30 million annual metric tons of CO_2-equivalent greenhouse gases . . . than those of the 63 lowest-emitting countries combined (Rafey and Sovacool 2011: 1141).

Second, what have been termed "unconventional" fossil fuels, have rapidly become a central part of a new discourse in which fossil fuels can play a key role, perhaps controversially, in a low-carbon transition. While techniques for extracting natural gas from coal or shale, now known collectively as 'fracking', were developed in the US in the 1940s, its recent upsurge has been driven by the discovery of new reserves and the adaptation of techniques used in deepwater oil and gas wells and the political imperatives of energy independence in the US (Davis 2012: 179). The vision of energy independence has been a powerful driver for an energy transition, particularly during a period at which the geopolitical struggles with which the US engaged have been associated with securing fossil fuel resources. Charles Davis reports that a

> study by IHS Cambridge Energy Research Associates (IHS Global Insight, 2010) indicated that shale gas released from unconventional sources amounted to 1 percent of gas supplies in 2000; however, this figure had increased to 20 percent by 2010 and is expected to reach 50 percent by the year 2035.
>
> (ibid.)

The transition to natural gas energy systems has by no means been confined to the US, with new developments emerging across Europe, notably in Central and Eastern European countries where dependence on imported fuel from Russia is a significant political issue. In "response to the shale gas boom in the US, the UK has seen the emergence of a nascent shale gas industry" supported by the development of scientific assessments of the extent and nature of shale gas resources, investments from a new set of industry actors, and a favourable policy climate (Cotton et al. 2014: 427). In 2013 the Conservative–Liberal coalition government developed a policy framework including tax incentives for industry, a more permissive regulatory framework, benefit packages for com–munities that would host shale gas extraction, and "a 100% business recovery (rate) from fracking operations for local authorities . . . resulting in an estimated 1.7 million per annum for a typical shale gas site funded by central government" (Cotton et al. 2014: 428). Research found that advocates of the development of shale gas in the UK – such as policymakers at the Department for Energy and Climate Change, and representatives of the industries and consultants involved in the nascent business of extracting shale gas – often explicitly considered shale gas to be part of a "transition" from existing energy systems to a more desirable future. Matthew Cotton (2014: 432) and his colleagues found that

> these actors universally adopted the language of "bridges" and "pathways," implying a process of "visible and coherent transition management" which recognised the UK's continued economic dependence on fossil fuels but which reduced GHG emissions when compared to coal or oil.

In contrast, those opposed to the development of shale gas reserves, such as environmentalist groups, tended to frame this vision as a 'distraction', or a pathway that would not lead to a low-carbon energy future but would rather sustain existing dependence on fossil fuels. The emergence in the UK of explicit interventions designed to translate the vision of a fossil fuel-based transition to a more secure and sustainable energy future alongside attempts to enable a more direct transition to low-carbon economies – through renewable energy and energy demand management, for example – demonstrates the ways in which visions of energy futures can co-exist, with some advocates finding complementarity while others see competition or outright conflict between different pathways. In the US, too, researchers have found that the "discursive framing of natural gas as a green fossil fuel" together with its potential as "a solution for national resource independence and domestic energy needs, and a generator of local economic growth" has served as a means through which fracking has come to be framed as an essential component of any transition to a stable energy future (Finewood and Stroup 2012: 73).

Reduced demand

While narratives that articulate the need for a low-carbon transition aim to foster a distinctive break with existing energy systems through shifting the basis through which energy is produced, the emergence of the fossil fuel-based narrative of gas as a "bridging fuel" towards low-carbon energy futures demon-strates the ways in which such discourses do not fundamentally challenge the link between social well-being, economic growth and the continued expan-sion of energy resources. In contrast, some visions of energy futures depict an altogether different future, in which energy use is not only changed to different resources, but this is accompanied by reduced demand. The need to conserve energy and resources has long been part of the discourse of modern society, becoming particularly prominent at times of national crisis or energy shortage, for example during periods of war or the oil crises of the 1970s. While the idea of energy efficiency has therefore taken hold across policy and public discourses around energy systems, more radical ideas of conservation have only been widespread during these times of emergency. Nonetheless, narratives concerning the need for radical reductions in energy demand have persisted as one means through which future visions of energy systems are articulated. In the 1980s, Amory Lovins of the Rocky Mountain Institute advanced the *Factor 4* argument that similar levels of economic productivity and social well-being could be achieved using one quarter of the energy. In more recent work, the Rocky Mountain Institute has sought to advance these arguments through their campaign, *Reinventing Fire*. In this campaign they advocate the development of a range of interventions that would enable economic growth in the US while maintaining or reducing current energy levels, through investments which "must earn at least a 12%/y real return in industry, 7% in buildings, and 5.7% in electricity, and new autos must repay any higher price within three years" (Rocky Mountain Institute n.d.). The key argument is that such up-front investment, while appearing costly, has the potential to save significant resource and finance over the short and long term, while also addressing concerns for energy security and affordability that have dominated the continued entrenchment of fossil-fuel based energy systems.

Alongside such advocacy groups, discourses of energy reduction have perhaps found their most visible expression amongst grassroots groups–activists and communities seeking to develop alternative ways of living that are less dependent on energy consumption in general and fossil fuels in particular. A prominent example of such action is the Transition movement. Originating in Ireland in 2005 and adopted as a vision for the energy future of Totnes, Devon, UK, in 2006, the Transition movement advocates the localisation of economies such that they are less dependent on fossil fuels and reduce their overall consumption of energy. As the Transition Network (2018), an organisation established to support the 1,000+ Transition Initiatives that can now be found globally suggests, this requires a different vision of the energy system than the one that currently dominates:

> All industrialised countries appear to operate on the assumption that our high levels of energy consumption, our high carbon emissions and our massive environmental impact can go on indefinitely. And most developing countries appear to aspire to these ways of living too. However, any rational examination of our energy supplies, our economic inequalities, our diminishing levels of well-being, our ecological crises and the climate chaos that is already hitting millions of people tells us this can't go on

much longer. We're saying that the best place to start transitioning away from this unviable way of living is right within our own communities, and the best time is right now.

Addressing these challenges is seen to require what some term 'post-carbon' energy systems, no longer dependent on fossil fuels. The LILAC (Low Impact Living Affordable Community) co-housing community in Leeds sought to put this approach into practice through the development of a 0.7 ha site in the city. Central to its philosophy was the idea of building very low impact and energy efficient housing, and for this purpose "straw and timber were preferred as they provided opportunities for elements of community self-build but also because they can be sourced locally and create value added local supply chains" (Chatterton 2013: 1660). In addition to on-site renewable energy generation, the development adopted a "collective design methodology ... to design out certain carbon intensive elements. The overall site was designed with a shared common house and car-free home zone in the middle. Car restraint and segregation were principle components" (Chatterton 2013: 1661). Another example of where visions of "post-carbon" living are being put into practice is through grassroots groups, such as Carbon Reduction Action Groups (CRAGs). One example is in the village of Fawnhope, Herfordshire, UK, where the community group was established to "support each other in reducing [carbon] footprints, sharing skills and knowledge in lower carbon living and promoting awareness and practical action in the wider community." (https://sites.google.com/site/350treesproject/Home).

TRANSFORMING ENERGY SYSTEMS?

The discursive and visual representation of different visions of energy futures plays an important role in structuring the ways in which existing energy systems are organised, reconfigured and challenged. Yet, changing the basis of the ways in which contemporary societies currently generate and use energy requires further forms of intervention through which these visions and narratives can be realised and, often literally, set in stone. Social scientists have traditionally focused on understanding the ways in which changes to policy – such as regulations, financial incentives, or targets – shape the development and trajectory of social and environmental issues such as energy. As set out in the first half of this chapter, such forms of policy intervention have been critical in giving substance to narratives of different energy futures, often in quite conflicting ways. However, the socio-technical nature of energy systems and their embedded nature means that change requires not only forms of social intervention, but other forms of innovation and intermediation which are able to reconfigure the material and technical components of energy systems in line with new social norms, institutions, and everyday practices. The need to understand the socio-technical dynamics of energy system change has been one driver behind the growing interest in studying transitions. In **Chapter 9**, we saw how approaches have been developed to analyse historical shifts in the provision and use of energy and introduced the *multilevel perspective* on energy systems in transition. The multilevel perspective examines how such shifts take place through looking at how 'niche' innovations come to be aligned with pressure on incumbent regimes. In this model, the development of niche innovations is seen as a particularly important component for achieving system change. Researchers have therefore also sought to understand whether and how purposeful attempts to develop new forms of innovation might be able to catalyse broader shifts in energy systems.

One means by which such transitions are thought to be fostered is through what has been termed strategic niche management (Raven et al. 2016). Within a socio-technical system, niches are seen to operate at the micro-level of the system and to provide "protection" against the prevailing technologies and interests of the existing regime, such that the

> development of sustainable technologies can take place and supportive constituencies can be built. Niche protective spaces shield the innovation against premature rejection by incumbent regime selection pressures, until the innovation is proven to be sufficiently robust to compete and prosper in unprotected market settings.
>
> (Smith et al. 2013: 2)

Niches are seen to work by enabling the stabilization of innovations, whether these are new energy generation technologies or social innovations, such as new forms of financing or owning energy systems. Stabilizing niches requires both the growth and development of the innovation and the social networks that sustain it and forms of learning, so that common understandings and narratives are developed amongst the actors involved (Brown & Vergragt 2008). Creating these conditions has often been regarded as the work of central governments, through the use of subsidies or specific forms of regulation, such as feed-in-tariffs that provide incentives for the development of micro-scale renewable energy. There are, however, a growing number of organisations and groups involved in the development of such niche innovations. Researchers have documented the growth of **grassroots innovations**, supported and sustained by community groups (Smith et al. 2016). Private sector corporations have invested in various demonstration projects to showcase technologies or to provide models of alternative forms of energy generation and use. One example is the development by the European utility company E.ON of a block of seven apartments in Malmo, Sweden, designed to showcase different forms of energy generation technology alongside 'smart' monitoring and data

grassroots innovations
a specific form of niche innovation developed by local or community groups, usually originating from outside mainstream political and economic processes, though often with the support of external funding.

Figure 10.3 An urban laboratory for smart living in the making? Little Kelham, Sheffield, UK (photo credit: Harriet Bulkeley)

visualisation technologies that allow householders to track their energy use and respond to different price signals. Such demonstration projects are now being integrated into the mainstream as urban redevelopments promise future residents 'energy smart' ways of living. Figure 10.3 provides an illustration of one such development being built in the former steel working town of Sheffield, UK.

The development and emergence of multiple forms of niches aimed at fostering new interventions through which energy futures cannot only be imagined but put into practice raises questions about both the geographies and politics of transitions. On the one hand, it suggests that thinking of socio-technical regimes as operating at the national level may miss the complexity of the multiple systems of energy provision and use that co-exist and the different arenas through which innovation is taking place. On the other hand, the presence of alternative visions of energy futures coupled with a diverse array of actors and institutions engaged in the development of niches and experiments for energy system change suggest that there are likely to be significant forms of contestation and conflict within and between different regimes and niches as attempts are made to realise different forms of energy futures. In the rest of this section, we explore some of these issues through examining the ways in which niches and experiments are being developed in urban arenas in order to realise visions of low-carbon energy futures.

Forging low-carbon urban energy futures

Research has shown that cities play an important role in the geographies of transitions (Coenen et al., 2012; Truffer & Coenen, 2012; Hodson & Marvin, 2010). Geels (2010) identifies two main roles that cities might play in socio-technical transitions: first, as an actor, where municipal actors may lead transitions; second, as a theatre for action, such that the urban arena provides spaces of innovation that can act as seedbeds for transitions. The growing interest in the development of Urban Living Laboratories suggests that cities are regarded as capable of fulfilling both of these roles in the formation of niches within energy systems (Evans & Karvonen 2014).

Yet, there are a growing range of forms of intervention and experimentation that exceed these roles, given the multiple networks and circulations through which urban economies and societies are constituted, such that they are seen as central to the orchestration of broader forms of experimentation within energy regimes. Hughes' (1983) analysis of the history of electrification drew attention to the ways in which such socio-technical systems were contextually, geographically and historically produced such that some cities were at the centre of the transition while others were bypassed (Bulkeley et al. 2015). Likewise, we might imagine contemporary processes of energy system reconfiguration to involve a more or less central role for some (parts of) cities than others. For Monstadt (2009: 1937), this means that we need to examine transitions not only through national energy regimes, but through the urban infrastructure regimes, "stable urban configurations of institutions, techniques, and artefacts which determine 'normal' sociotechnical developments in a city and thus shape general urban processes and the urban metabolism" through which energy futures are being forged and contested.

In their analysis of the development of urban experiments focused on climate change, Harriet Bulkeley and colleagues (2015: 22) find that

> up to 45 per cent of the ... [627] experiments [recorded in a survey of one hundred global cities] seek to intervene in energy systems, and include efforts aimed at reducing demand as well as new systems of energy production and generation in the urban infrastructure sector.

These urban climate change experiments take a variety of forms. Some are focused on the development of new forms of energy technology, while others are concerned with forms of social innovation. In Berlin, for example, municipal actors sought to develop a map of the solar potential of rooftops across the city to provide the basis for private sector investment in the technology, while in Philadelphia a public private partnership between the municipality and Dow Chemicals led to a scheme for communities to enter a competition to win a 'cool roof' that would improve the energy efficiency of a residential block (Bulkeley et al. 2014). There are also examples of community-led experiments in cities. In Brixton, London, community groups involved with the Transition

---

EXAMPLE 2: SMART GRID SMART CITY

Australia's first commercial-scale smart grid project – *Smart Grid Smart City* – took place in the Newcastle–Sydney region of New South Wales between 2010 and 2014. Smart grids, with their capacity to manage real-time demand for electricity in relation to available supply, are seen as central to managing peak energy demand more efficiently and a means for enabling the integration of renewable energy generation in electricity provision. Given the density of electricity use in urban areas, peak demand can be a particular challenge for cities and one that may become more intense as cities experience hotter summers and the growing use of electrically powered air conditioning. At the same time, cities provide an important site where managing and reducing demand can have a significant effect on overall consumption of electricity, and related greenhouse gas emissions. In Australia, cities are also important in the generation of renewable energy, with more than one in five Australian households now having solar photovoltaic installations.

Smart Grid, Smart City deployed 25,000 advanced smart meters providing real-time information about energy use, and groups of households took part in trials of different means through which they could reduce their demand at times of peak use. In addition, 60 consumer batteries, 25 fuel cells and 10 small wind turbines were installed, and network monitoring and fault-detection technologies were used to improve the grid. Customers involved in the trial were given different tariffs, reflecting the changing prices of electricity seasonally or in real-time, in order to see whether they would change their behaviour in line with the cost of electricity.

At the end of the AUS$100 million project, the Australian government concluded that "adoption of smart grid technologies across the National Electricity Market would deliver significant economy wide benefits in the order of $9.5–$28.5 billion over 20 years" (Government of Australia 2017). Yet, the insertion of smart grids into the urban economy is not straightforward. Multiple technical challenges were encountered in the project, from mundane matters (such as whether or not the signal to the smart meter could penetrate through different kinds of walls), to wider concerns about who can benefit from the installation of smart meters and flexible tariffs (Bulkeley et al. 2016). Smart grids are not only being installed in cities but, at the same time, are actively shaping the urban arena – creating new energy landscapes, changing household economies and new forms of everyday practice.

Town movement have leveraged funding and opportunities provided by a Low Carbon Zone established by the Mayor of London. The community has developed new schemes for improving the energy efficiency of homes, and established an energy co-operative through which investment in solar power can provide renewable electricity within the community.

The range and extent of experimentation taking place in cities in response to the vision of a low-carbon energy future points to the diversity of actors and interests involved, and the potential for contradiction and conflict between the different goals and dynamics being pursued. As the first part of this chapter suggested, there is no fixed definition of a low-carbon energy system, with interpretations ranging from those based on new forms of fossil fuel to those taht are powered by renewables, from those in which existing patterns of consumption are maintained to those in which radical reductions in energy use are required. Rather than identifying particular niches that might be more successful than others in shifting the existing regime, this complexity and its politics suggests that there are likely to be multiple forms of transition at work in any one context. In particular, it is important to recognise that the nature, scope and pace of change will vary between different sectors and geographical contexts, depending on historical patterns of energy system development (**Chapter 9**) as well as the geopolitical, economic, and social imperatives shaping energy provision and use. The potential for energy transitions is, therefore, not only shaped by the availability of resources (e.g. the solar potential) but of the conditions within which such transitions take place, and the ways in which they are enabled and contested (e.g. legislative change to enable the integration of solar panels with buildings, economic incentives or feed-in tariffs for solar power, daily patterns of electricity use that enable solar power to be productive, the availability of storage etc., opposition to the loss of aesthetic value in streetscapes with solar panels or resistance from incumbent energy providers).

Yvonne Rydin and her colleagues (2013) suggest that a **pathways approach** can enable an analysis of the contested nature of energy system change. Rather than viewing transitions as a change from one regime to another, this approach focuses on transitions as "the outcome of ongoing interactions between technologies, political and economic frameworks, and between institutions and social practices, during which these different dimensions change or coevolve together to produce distinct pathways of change" (Rydin et al. 2013: 637). Melissa Leach and her team (2010: 48) also suggest that a pathways approach starts from the assumption that "different actors and networks, framing systems dynamics, boundaries and goals in different ways, produce very different narratives about what a response should be and what might make it effective" in terms of realising sustainability. Pathways to energy futures are not linear and always contested. There are multiple pathways open and others that remain more hidden from view. Understanding and opening up these pathways will be crucial to charting any future energy transition.

pathways approach
rather than viewing energy transitions as inevitable, linear processes, a pathways approach encourages us to think of transitions as involving many different junctures, each of which offers different routes through the landscape of existing energy systems and which allows us to think of multiple possible energy futures.

CONCLUSIONS: REALISING ENERGY FUTURES?

Energy has long been taken for granted as the backdrop against which society operates. Often considered primarily as a matter for the engineering sciences, its important social, economic and political dimensions have either been overlooked or considered as singular issues. However, as a host of issues, from air pollution to resource conflicts, access to energy for development to climate change have placed energy firmly on the public agenda, so too have social scientists come to understand energy as thoroughly entwined with the modern social world. In particular, the growing recognition of the gap between the ambitious targets set in the 2015 Paris Accord for reductions in greenhouse gas emissions compatible with keeping temperature rises within 1.5 degrees globally, and the international agreement of Sustainable Development Goals that seek to provide access to clean energy services, mean that energy transitions are no longer simply of historical interest – they are of pressing concern world-wide.

Recognising that energy is not a stand-alone sector, this book has offered a critical social science perspective on energy systems that highlights the complex interplay of resource, economic, infrastructural and geopolitical landscapes (**Chapters 1–4**). Rather than being determined by the 'natural' abundance of resources (e.g. of gas for fracking, coal for mining, or solar radiation) and the technical capacity to access these resources, patterns of energy production and consumption are a result of, for example, the interweaving of cultural norms surrounding how and for what energy is used, investment decisions by multinational companies, regulation and policy goals, environmental concerns, social mobilisation, patterns of industrial growth and decline and so forth. At the same time, these energy geographies are critical to the reproduction of social structures and everyday life. Energy provision and use is ingrained in state territorialisation, the ways in which cities and built and inhabited, systems of mobility, practices of cooking and norms of office clothing. Energy configures and reproduces social, economic and political geographies, while at the same time these geographies create and sustain particular energy systems.

The landscapes of energy provision and use that we see before us today are, therefore, a result of the specific ways in which these processes have come together under particular conditions over time. In their broadest sense, encompassing the physical, material 'hard wiring' of the energy system as well the institutions, practices, norms, cultures and actors that are bundled together, energy infrastructures can 'lock in' particular energy geographies. Sedimented forms of economic production and consumption similarly tend to reproduce existing patterns of (uneven) economic development, while energy's complex geopolitics reinforces circulations of capital and power on the global stage that shape energy futures in all too familiar directions.

Yet, as the chapters in this book have shown, energy systems are not static but continually evolving and open to change. Three dynamics appear to be particularly important in shaping how energy futures emerge and take hold. The first is that of *development*. Social and economic development continue to shape the uneven landscapes of energy provision, access and use. As societies have become more affluent and their energy demands more complex, so too has demand for energy been built into the structure of the state, geopolitical relations between different countries and the ways in which economic and cultural norms evolve. At the same time, the uneven nature of economic development has led to energy poverty and vulnerability, manifest in different ways and to different degrees in the global South and the global North. Providing affordable, clean energy to all is now at the heart of the sustainable development agenda for the international community. Yet, whether and how such energy needs can be met while still sustaining the high levels of demand for energy use amongst wealthier communities and remaining within environmental limits remains a significant challenge.

The second is the dynamic of *innovation*. As the growing interest in energy transitions has demonstrated, how innovation emerges and the extent to which it is able to 'breakthrough' existing socio-technical regimes is crucial for the evolution of energy geographies. The history of energy systems is not only driven by the lock-in and sedimentation of existing political, economic, resource and infrastructural landscapes, but also one characterised by technical breakthroughs and social change. Within the energy arena, innovation has frequently emerged in response to perceived problems and challenges – whether this be in the form of air pollution, matters of energy security, access, affordability or most recently climate change. Yet, the pathway from emerging niches, where alternative forms of energy provision, ownership, organisation and use are being tried out to widespread uptake and embedding within national and international energy systems is far from smooth.

While development and innovation are critical in shaping energy futures, it is perhaps the third dynamic, of *contestation*, that is most significant in determining how future energy geographies are shaped. As we have seen in this book, energy attracts a great deal of controversy and conflict. From specific disputes about the development of particular energy resources, systems or infrastructures, to concerns about the uneven distribution of the costs and benefits of our forms of energy provision for society and for the environment, energy is being contested. Contestation can take the form of international political dispute, over securing access to new energy resources for example. Or it can emerge within particular national contexts as different social and economic interests align around different possible energy futures. Contestation also arises at the local scale, as communities seek to gain access to energy provision or insert new ways of producing and owning energy into incumbent systems. Whether conducted through highly visible protests or through quieter registers of resistance, contestation is vital in shaping how the dynamics of development and innovation gain momentum to realise energy futures.

Realising new energy futures that can address the four imperatives of access, security, sustainability and social justice cannot then be a matter of identifying a single technical silver bullet or of hoping that if we are able to produce enough clean energy issues concerning uneven development, affluent over-consumption and environmental justice will readily be wished away. Rather, what lies ahead are multiple pathways, each of which is shaped by dynamics of development, innovation and contestation and none of which can, by themselves, offer a perfect solution. A critical social science understanding of the pathways, challenges and choices ahead provides a compass to offer direction towards the destinations our future energy systems must find.

QUESTIONS FOR DISCUSSION

- In what ways does it make sense to talk about energy transitions as a 'political project'?

- What are the similarities and differences between different ideas of energy transition?

- How have social scientists tried to explain and model the process of transition? What are the advantages and limitations of such approaches?

- To what extent is innovation or experimentation central to the process of transition, and what are the implications for how we understand the politics of transition?

- Who wins, and who loses, from energy transitions?

ACTIVITIES/POTENTIAL RESEARCH PROJECTS

- For a city with which you are familiar, develop two energy transition plans – how would you seek to develop and implement a low-arbon transition, and what would a high-carbon transition entail? Which scenario do you think is most likely and why? Compare your approach with current policy documents or plans for the city.

- Examine the low-carbon transition plan for a national government – how is the idea of a low-carbon transition being framed, what benefits is it thought to produce, for whom, and how, where and when will it be implemented? If you are working in a group, divide this task between you so that multiple plans are analysed and then hold a class discussion to debate the similarities and differences.

RECOMMENDED READINGS

■ Baker, Lucy, Peter Newell and Jon Phillips 2014. The political economy of energy transitions: the case of South Africa. *New Political Economy* 19 (6): 791–818.

Explores the contested politics of energy transitions in the context of South Africa. The paper demonstrates the importance of examining the politics of energy transitions and how energy landscapes are shaped by resources, infrastructures, economies and geopolitics.

■   Bulkeley, H., V. Castan Broto and G. Edwards. 2015. *An urban politics of climate change: experimentation and the governing of socio-technical transitions*, London and New York: Routledge.

This book examines the role that cities are playing in low-carbon transitions. It establishes the emergence of the low-carbon agenda at the urban level and suggests that 'experimentation' has become the dominant mode through which diverse urban actors are seeking to respond to this agenda. It contains case-study chapters from cities in the global North and global South through which these issues are explored and illustrated.

■   Rydin, Y., C. Turcu, S. Guy and P. Austin. 2013. Mapping the coevolution of urban energy systems: pathways of change. *Environment and Planning A* 45(3): 634–649.

Through an analysis of urban energy initiatives within the UK, this paper examines how energy systems co-evolve through technical and social change and at the multiple pathways though which new modes of energy provision and use are being developed.

■   Smith, A., J.P. Voß and J. Grin. 2010. Innovation studies and sustainability transitions: the allure of the multi-level perspective and its challenges. *Research Policy* 39: 435–448.

A substantial review of the concept of socio-technical transitions that provides an excellent summary and point of reference through which to understand this approach. The article focuses on the use of the multi-level perspective, giving clear explanations of its strengths and limitations.

REFERENCES

Brown, H.S. and P.J. Vergragt. 2008. Bounded socio-technical experiments as agents of systemic change: the case of a zero-energy residential building. *Technological Forecasting and Social Change* 75(1): 107–130.

Bulkeley, H., V. Castan Broto and G. Edwards. 2015. *An urban politics of climate change: experimentation and the governing of socio-technical transitions*, London and Bew York: Routledge

Bulkeley, H. and Newell, P. 2015. *Governing climate change*, Abingdon and New York: Routledge. Second edition.

Bulkeley, H., V. Castan Broto, M. Hodson and S. Marvin (eds). 2010. *Cities and low carbon transitions*. London: Routledge.

Bulkeley, H.A., P. McGuirk and R. Dowling. 2016. Making a smart city for the smart grid? The urban material politics of actualising smart electricity networks. *Environment and Planning A* 48: 1709–1726.

Chatterton, P. 2013. Towards an agenda for post-carbon cities: lessons from LILAC, the UK's first ecological, affordable, cohousing community. *International Journal of Urban and Regional Research* 37(5): 1654–1674.

Clinton Foundation. 2008. Press release: First Clinton global initiative commitment of 2008 | Clinton Foundation. Retrieved 28 October 2013, from http://www.clintonfoundation.org/main/news-and-media/press-releases-and-statements/press-release-first-clinton-global-initiative-commitment-of-2008.html

Coenen, L., P. Benneworth and B. Truffer. 2012. Toward a spatial perspective on sustainability transitions. *Research Policy* 41(6): 968–979.

Cotton, M., I. Rattle and J. VanAlstine. 2014. Shale gas policy in the United Kingdom: an argumentative discourse analysis. *Energy Policy* 73: 427–438.

Davis, C. 2012. The politics of "fracking": regulating natural gas drilling practices in Colorado and Texas. *Review of Policy Research* 29(2): 177–191.

DTI. 2003. *Our energy future – creating a low carbon economy: Energy White Paper.* London The Department of Trade and Industry.

Evans, J.P. and A. Karvonen. 2014. 'Give me a laboratory and I will lower your carbon footprint!' – urban laboratories and the governance of low-carbon futures. *International Journal of Urban and Regional Research* 38 (2): 413–30

Finewood, M. and Stroup, L. 2012. Fracking and the neoliberalization of the hydro-social cycle in Pennsylvania's Marcellus Shale. *Journal of Contemporary Water Research & Education* 147: 72–79.

Geels, F. 2010. The role of cities in technological transitions. In *Cities and low carbon transitions.* H. Bulkeley, V. Castán Broto, M. Modson and S. Marvin (eds). London: Routledge, pp. 13–28.

Government of Australia. 2017. Smart grid, smart city. Available online at www.environment.gov.au/energy/programs/smartgridsmartcity

Hodson, M. and S. Marvin. 2010. Can cities shape socio-technical transitions and how would we know if they were? *Research Policy* 39(4): 477–485.

Hughes, T. 1983. *Networks of power electrification in western society, 1880–1930.* Baltimore, MD: John Hopkins University Press.

Leach, M., I. Scoones and A. Stirling. 2010. *Dynamic sustainabilities: technology, environment, social justice.* London: Earthscan.

Lovell, H., H. Bulkeley and S. Owens. 2009. Converging agendas? Energy and climate change policies in the UK. *Environment and Planning C* 27(1): 90–109.

Monstadt, J. 2009. Conceptualizing the political ecology of urban infrastructures: insights from technology and urban studies. *Environment and Planning A* 41(8): 1924–1942.

Moss, T., S. Becker and M. Naumann. 2015. Whose energy transition is it, anyway? Organisation and ownership of the Energiewende in villages, cities and regions. *Local Environment: The International Journal of Justice and Sustainability*, 20 (12): 1547–1563.

PIU 2002. *The energy review.* Performance and Innovation Unit, Cabinet Office, London: HM Government.

Rafey, W. and B. K. Sovacool. 2011. Competing discourses of energy development: the implications of the Medupi coal-fired power plant in South Africa. *Global Environmental Change* 21: 1141–1151.

Raven, Rob, Florian Kern, Bram Verhees and Adrian Smith. 2016. Niche construction and empowerment through socio-political work. A meta-analysis of six low-carbon technology cases. *Environmental Innovation and Societal Transitions* 18: 164–180.

Rocky Mountain Institute. (n.d.) Available online at http://www.rmi.org/reinventingfire

Rydin, Y., C. Turcu, S. Guy and P.Austin. 2013. Mapping the coevolution of urban energy systems: pathways of change. *Environment and Planning A* 45(3): 634–649.

Smith A., T. Hargreaves, S. Hielscher, M. Martiskainen and G. Seyfang. 2016. Making the most of community energies: three perspectives on grassroots innovation. *Environment and Planning A* 48(2): 407–432.

Smith, A., F. Kern, R. Raven and B. Verhees. 2013. Spaces for sustainable innovation: solar photovoltaic electricity in the UK. *Technological Forecasting and Social Change* 81: 115–130.

Stegen, K.S. and M. Seel, 2013. The winds of change: how wind firms assess Germany's energy transition. *Energy Policy* 61: 1481–1489.

TERI. (n.d.). Lighting a billion lives. Retrieved 08 September 2014, from http://labl. teriin.org/index.php?option=com_dir&task=all

Transition Network. 2018. Transition towns movement, great transitions stories. Available online at www.greattransitionstories.org/wiki/Transition_Towns_Movement

Truffer, B. and L.Coenen. 2012. Environmental innovation and sustainability transitions in regional studies. *Regional Studies* 46(1): 1–21.

UN. 2018. Low carbon development. Sustainable Development Knowledge Platform. Available online at https://sustainabledevelopment.un.org/index.php?menu= 1448#

UNFCC. 2009. Copenhagen Accord. United Nations. Available online at http://unfccc. int/meetings/copenhagen_dec_2009/items/5262.php

UNFCC. 2011. Cancun Agreements. United Nations. Available online at http://unfccc. int/cancun/

Index

Note: Indicators in **bold** refer to tables; those in *italic* refer to figures.

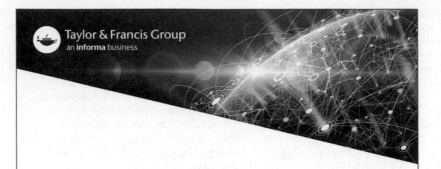